T0397335

A Stitch in Line

A Stitch in Line: Mathematics and One-Stitch Sashiko provides readers with instructions for creating hitomezashi items with minimum outlay. The reader is guided through the practical steps involved in creating each design, and then the mathematics which underpins it is explained in a friendly, accessible way. This is a fantastic book for anyone who is interested in recreational mathematics and/or fibre arts and can be a useful resource for teaching and learning mathematical concepts in a fun and engaging format.

Features

- Numerous full-colour photographs of hitomezashi stitch patterns which have been mathematically designed.
- Suitable for readers of all mathematical levels and backgrounds — no prior knowledge is automatically assumed.
- A compressed encoding for recording and designing hitomezashi patterns to be stitched or drawn.
- Accessible explanations and explorations of mathematical concepts inherent in, or illustrated by, hitomezashi patterns.

Katherine Seaton holds a B. Sc. and Ph. D. in mathematics from the University of Melbourne and a Graduate Certificate in Higher Education from La Trobe University, both in Australia. For more than 27 years she worked in teaching and research in the Department of Mathematics and Statistics at La Trobe where, now retired, she holds an honorary position. Katherine's original research was in mathematical physics; over time she added tertiary mathematics education (assessment and academic integrity) and mathematical art to her areas of interest. As well as her formal educational qualifications, she learned to stitch, knit and crochet at a young age, from her mother, her grandmother and her Brownie Guide leaders respectively. Katherine is a strong advocate for using fibre arts to widen perspectives on what it means to do mathematics and on who is a mathematician. She has published in the Journal of Mathematics and the Arts and has spoken and exhibited hitomezashi at Bridges conferences since 2018. Where possible, she uses repurposed materials in her practice. Her mathematical fibre art can be found on instagram: @maths_kath.

AK Peters/CRC Recreational Mathematics Series

Series Editors

Robert Fathauer
Snezana Lawrence
Jun Mitani
Colm Mulcahy
Peter Winkler
Carolyn Yackel

Mathematical Conundrums
Barry R. Clarke

Lateral Solutions to Mathematical Problems
Des MacHale

Basic Gambling Mathematics
The Numbers Behind the Neon, Second Edition
Mark Bollman

Design Techniques for Origami Tessellations
Yohei Yamamoto, Jun Mitani

Mathematicians Playing Games
Jon-Lark Kim

Electronic String Art
Rhythmic Mathematics
Steve Erfle

Playing with Infinity
Turtles, Patterns, and Pictures
Hans Zantema

Parabolic Problems
60 Years of Mathematical Puzzles in Parabola
David Angell and Thomas Britz

Mathematical Puzzles
Revised Edition
Peter Winkler

Mathematics of Tabletop Games
Aaron Montgomery

Puzzle and Proof
A Decade of Problems from the Utah Math Olympiad
Samuel Dittmer, Hiram Golze, Grant Molnar, and Caleb Stanford

A Stitch in Line
Mathematics and One-Stitch Sashiko
Katherine Seaton

The Four Corners of Mathematics
A Brief History, from Pythagoras to Perelman
Thomas Waters

For more information about this series please visit: https://www.routledge.com/AK-Peters CRC-Recreational-Mathematics-Series/book-series/RECMATH?pd=published,forthcoming &pg=2&pp=12&so=pub&view=list

A Stitch in Line

Mathematics and One-Stitch Sashiko

Katherine Seaton

CRC Press is an imprint of the
Taylor & Francis Group, an **informa** business
AN A K PETERS BOOK

Designed cover image: Katherine Seaton

First edition published 2025
by CRC Press
2385 NW Executive Center Drive, Suite 320, Boca Raton FL 33431

and by CRC Press
4 Park Square, Milton Park, Abingdon, Oxon, OX14 4RN

CRC Press is an imprint of Taylor & Francis Group, LLC

Library of Congress Cataloging-in-Publication Data
Names: Seaton, Katherine, author.
Title: A stitch in line : mathematics and one-stitch sashiko / Katherine Seaton.
Description: First edition. | Boca Raton : AK Peters/CRC Press, 2025. | Series: AK Peters/CRC recreational mathematics series | Includes bibliographical references and index.
Identifiers: LCCN 2024028720 (print) | LCCN 2024028721 (ebook) | ISBN 9781032491509 (hbk) | ISBN 9781032487205 (pbk) | ISBN 9781003392354 (ebk)
Subjects: LCSH: Sashiko--Mathematics. | Quilting--Patterns. | Mathematical recreations.
Classification: LCC TT835 .S43135 2025 (print) | LCC TT835 (ebook) | DDC 746.46--dc23/eng/20240807
LC record available at https://lccn.loc.gov/2024028720
LC ebook record available at https://lccn.loc.gov/2024028721

ISBN: 978-1-032-49150-9 (hbk)
ISBN: 978-1-032-48720-5 (pbk)
ISBN: 978-1-003-39235-4 (ebk)

DOI: 10.1201/ 9781003392354

Typeset in Latin Modern
by KnowledgeWorks Global Ltd.

Publisher's note: This book has been prepared from camera-ready copy provided by the authors.

"each stitch a deliberate thought or small prayer"

Carol Hayes (1962–2022)
colleague and friend

Contents

Preface

THIS book is all about hitomezashi and the mathematics inherent in it, and that can be experienced by making it. So you might wonder why the word 'hitomezashi' doesn't appear explicitly in the title. It's there in partial disguise as 'one-stitch sashiko', which is what it means. When I told people what I was working on, far more of my friends and colleagues recognised the word 'sashiko' than did 'hitomezashi'. And it's probably down to me, but there is also another whole group of people (most of whom I have never met in person) who think that hitomezashi is a kind of mathematical drawing. A book's title needs to speak clearly of its subject matter to the people who might open it.

I began stitching small pieces of hitomezashi in 2019, investigating it as a potential form of mathematical fibre art to be added to a repertoire of interactive mathematical craft activities used with young and old in outreach. We hadn't done quite so much pivoting then as we have since, but reflecting on a past experience of mathematical embroidery, needles and public participation—although it's not the needles one can see so much as the missing needles that one has to worry about—sent me looking for a safer way for large groups to create the entrancing patterns of hitomezashi. Drawing on a grid was a clear alternative, but it needed to be a more interesting and creative activity than simply copying a hitomezashi chart from one piece of paper to another.

Hitomezashi is comprised of lines of running stitch, worked in two directions. Encoding the two possible states of these lines as 0 or 1 provides an alternative to recording a stitching pattern as a chart. Finding new ways to choose the sequences of binary digits to stitch or draw gives much to notice and wonder about. My first three ideas were to determine them randomly using a coin toss, to use them to spell out messages and to use mathematical methods to specify them. Thus were the ideas found in Chapters 2, 5 and 7 born.

In the same year I came across a recent publication about sashiko, by fellow Australian Professor Carol Hayes. We met, and from her depth of knowledge as a scholar in Japanese studies, particularly cultural production, she opened the door to a new world for me, one that you meet in Chapter 1. The hitomezashi workshop that we had planned to conduct together in person at the Bridges Mathematical Art conference in 2020 took place not in Helsinki but online, from our respective homes.

In March of 2020, when schools all over the world closed their physical doors, mathematics teacher Annie Perkins started the 100 Day *Math Art Challenge*. She put out a call for activities that could be done with items that families were likely to have on hand. I had the paper-and-pencil hitomezashi activity thought out, with no expectation of being able to run it in person any time soon, and Annie took it to the world through her blog and on Twitter. Wonderful creations from families and virtual school classes were tweeted: drawn, stitched and coded. People asked really perceptive questions based on what they had noticed and wondered, some of which are answered in Chapters 2, 3 and 4. Katie Steckles wrote about the idea for *The Aperiodical*. Subsequently, in 2021 Ayliean McDonald made a video about the drawn form of hitomezashi for *Numberphile*—and this is probably why there is a whole group of mathematically inclined people who think that hitomezashi is a form of drawing. All but three of the pieces in this book can be either stitched or drawn; one must be drawn and two can only be stitched.

In Chapter 1, as well as learning about the origins of hitomezashi, you will find the practical instructions you need to get started. Chapter 2 introduces the method of stitching or drawing based on strings of 0s and 1s, and sets the pattern for all other chapters. Each begins with instructions for a sampler, a practical learning exercise. The mathematical idea introduced by making and observing that sampler is then explored in the rest of the chapter, and links made to other forms of art that relate to the same mathematical idea. Each chapter closes with some suggestions for more mathematics or stitching to explore. In some chapters, there may even be a second sampler to make. Chapters 2 and 3 discuss the most obvious thing that you will see in your hitomezashi: lots of loops, and loops inside loops. In Chapter 3, you will also meet the work of mathematicians who have never picked up a needle, so far as I know.

The next section of the book is called 'Three Big Ideas', ideas for which everyone, I hope, can get at least an intuitive feel. These ideas are respectively duality (Chapter 4), randomness (Chapter 5) and symmetry (Chapter 6). Hitomezashi joins the many other fibre arts which have had their potential symmetries explored from a mathematical viewpoint.

The final section of the book is called 'Generating Art'. In Chapter 7, we use hitomezashi to hide messages in plain view using ideas drawn from cryptography, and meet algorithmic art. Chapter 8 introduces what seem to be new designs determined mathematically. They have fractal properties, and specifying them involves the most technical mathematics (and the most equations) in the book. I hope you will find the designs as beautiful and as much reward-for-effort as I do. Every mathematics outreach program's favourite numbers (the Fibonacci numbers) make a guest appearance. In Chapter 9, known mathematical sequences with the property of quasiperiodicity are used as hitomezashi instructions. Another very obvious feature of hitomezashi is that the many loops are actually comprised of lines which make right-angled turns. Chapter 10 is called 'Corners' and the discussion ranges from the square lattice models of mathematical physics to the bathrooms of a museum (by way of Truchet tiles). Very recent research by other mathematicians is again mentioned in this chapter.

Up to this point, all the hitomezashi in the book is been drawn or stitched on a planar, square grid, in one colour. In Chapter 11, this is opened up to three dimensions and to triangular grids. You are invited to think about using more than one colour.

Something that has surprised and delighted me about the way the book has unfolded is that a fundamental idea in higher mathematics education, that of 'proof' has arisen very naturally. In the book, you will be introduced to proof by induction, existence proof by construction of an example, proof by cases (also called proof by exhaustion), visual proof, proof by contradiction, and direct proof. There is even an open conjecture presented, one awaiting proof.

I was not so surprised, because it has long been my belief, to realise how much mathematics is inherent in what were traditionally domestic arts, and how much tacit knowledge of mathematics is held by their practitioners. Here I don't mean measuring skills and being able to calculate stitch tension, though that is essential to making an object of the correct size. (With all due respect to the people who thought that is what I have been writing a book about, it would have been fairly short and probably a little bit boring.) Simple examples are the appreciation of positive and negative curvature necessary for a fitted garment, and the understanding of chirality that results in two wearable, rather than identical, gloves. I've filled a book with the mathematics in just one form of traditional embroidery and left you with more to explore.

During the writing of the book, I attended a virtual conference *Arts-based Research in Japanese Studies*, which encouraged me greatly and helped me make sense of where the writing was taking me. The keynote speaker (Vera Mackie) put forward the idea that there is not an absolute binary between research and creative outputs, and that one can 'triangulate' one's activities between reference points. I realised that in this book, I was triangulating between mathematics and Japanese fibre art. Conference co-convenor Megan Rose suggested that we can learn through artistic media things that cannot be put into formal language. (So don't just read, make the samplers.) The other co-convenor Sharon Elkind spoke of transcultural flows in fibre arts. In the second half of the twentieth century, American patchwork quilting, which in turn had its origins in Europe, was taken up in Japan. In a process of 'glocalisation' (the global made local), unique features such as the use of sophisticated taupe shades developed. Flows go both ways; sashiko has become known and practised outside Japan, as the explosion of recent books in English on the topic attests. In *this* book, you will find glocalisation of hitomezashi in the realm of mathematics.

Katherine Seaton
Melbourne
May 2024

About the Author

Katherine Seaton holds a B. Sc. and Ph. D. in mathematics from the University of Melbourne and a Graduate Certificate in Higher Education from La Trobe University, both in Australia. For more than 27 years she worked in teaching and research in the Department of Mathematics and Statistics at La Trobe where, now retired, she holds an honorary position. Katherine's original research was in mathematical physics; over time she added tertiary mathematics education (assessment and academic integrity) and mathematical art to her areas of interest. As well as her formal educational qualifications, she learned to stitch, knit and crochet at a young age, from her mother, her grandmother and her Brownie Guide leaders respectively. Katherine is a strong advocate for using fibre arts to widen perspectives on what it means to do mathematics and on who is a mathematician. She has published in the *Journal of Mathematics and the Arts* and has spoken and exhibited hitomezashi at Bridges conferences since 2018. Where possible, she uses repurposed materials in her practice. Her mathematical fibre art can be found on instagram: `@maths_kath`

Acknowledgements

I would like to thank the various people who have been very gracious in permitting images of their creation to be used in the book; their names appear with the images.

I am also very grateful to colleagues around the world for answering my emails about their work, or for suggesting articles or books that might be relevant: Abdalla Ahmed, Rodrigo Angelo, Michael Assis, sarah-marie belcastro, Julia Collins, Colin Defant, Jan de Gier, Noah Kravitz, Sébastien Labbé, Bernard Nienhuis, David Reimann, Madeleine Shepherd and Vincent Van Dongen.

Thanks also to Annie Perkins, Ayliean McDonald and Katie Steckles for putting mathematical hitomezashi out into the world.

The mathematical art community, many of who can be found at Bridges conferences and in the pages of the *Journal of Mathematics and the Arts*, is a warm and welcoming one.

Sometimes one simply has to stop stitching and 'shut up and write'. The various online writing groups by that name (and others) which I've joined over the last few years have pushed me along, as have the friends who've told me *not* to join them for coffee and who've checked that the writing has been happening, no matter how bemused they were by its subject matter.

I

Getting Started

CHAPTER 1

Background

SASHIKO emerged in domestic settings in rural Japan in the Edo period. This chapter provides a background to the development of sashiko and, in particular, introduces the one-stitch form *hitomezashi*. Practicalities for stitching the projects throughout the book are also covered.

1.1 THE EDO PERIOD: THE ORIGINS OF SASHIKO

In the Edo period (1603–1868 CE), the Tokugawa shogunate isolated Japan from the influence of the rest of the world. After many years of bitter civil war, this ushered in a period of peace, prosperity and stability, and a distinctively Japanese culture flourished. The poetry form *haiku*, once part of a longer poetry style, developed into an art in its own right in the Edo period. The seat of government Edo (present day Tokyo) grew to be a large city. In *ukiyo*, the floating or fleeting world, city dwellers could enjoy theatre, street food and other night-life. Woodblock prints *ukiyo-e* captured the essence of this world, including *kabuki* actors in their costumes and flamboyant fashion. Some prints show people engaged in *origami* paper-folding; paper became widely enough available during the Edo period that it could be used for recreation.

One master of ukiyo-e was Hokusai (1760–1849), famous for images of Mount Fuji such as *Under the Great Wave off Kanagawa*. He also published manuals of drawing and of design elements *Shingata Komoncho* (New Forms for Designs) for artisans; these tessellating, geometric elements endure in woodwork such as *kumiko* and *yosegi* (see Figure 1.1), as well as in sashiko and other crafts.

DOI: 10.1201/9781003392354-1

Figure 1.1 Contemporary yosegi tray showing design elements found in the manuals of Hokusai. Photograph by the author.

A distinctively Japanese mathematics *wusan* developed during the Edo period. There being no universities, scholars formed discussion groups led by a master. One such master Seki—a samurai during a time of peace—isolated from developments in European mathematics independently found what are generally called the Bernoulli numbers [Kitigawa, 2022]. Kate (Tomoko) Kitagawa further argues that beautiful book illustrations of this period show the emergence of abstraction and imagination, in contrast to concrete calculational facility. There were recreational and spiritual aspects to the practice of wusan. The solving of puzzles in geometry, many featuring circles, was a pastime for people of different ages and walks of life. Solutions would be inscribed in colour on wooden tablets *sangaku* to be hung in Shinto shrines and Buddhist temples. Whether an offering up of achievement or a visual encouragement for others to engage in wusan, they are also distinctive mathematical art [Rothman and Fukagawa, 1998, Fukagawa and Horibe, 2014].

The Edo period saw the rise of an urban merchant class. To curb ostentatious consumption and to maintain class distinctions and hence power, restrictive sumptuary laws were imposed by the shogunate. These laws prescribed the types of fabric and the patterns and colours that could be worn by commoners (farmers, artisans and merchants), the warrior class, and the ruling class respectively [Hayes, 2019]. Although theirs was regarded as a more honourable occupation than commerce, farming families in the cold northern regions were not permitted to wear cotton cloth, let alone silk. They fashioned their garments from the rough plant fibre hemp, grown, spun and woven at home, though dyed with indigo at a dyer's premises. When a family must manufacture every stitch they wear, grow their own food and generate produce to sell, every scrap of cloth and inch of thread is precious. Clothing was patched in layers, strengthened preemptively where wear was likely in daily work, and the rough open texture was softened and insulated with cotton stitching. Like a pair of loved jeans, indigo clothing became faded over time. It was passed from one generation to another, re-stitched and re-dyed as necessary. Woven rectangles that began as the panels or sleeves of a work jacket or trousers might next become part of a sleeping bag and eventually a dust-rag *zokin*.

Fishermen and firemen also wore padded, layered garments. The thick clothing of firemen, which included hoods and protection for the feet, would be doused with water so the wearer could bear the heat of a blaze. In a recent beautiful children's book *Sashiko* a young girl growing up on Awaji island in Japan's inland sea watches her mother devise warmer clothing for her fisherman father by layering together three garments into one, blue with white stitches [Ciletti and Pritelli, 2022]. She sees her mother's practicality and frugality, her nurturing spirit, good thoughts for the wearer in the long hours of stitching, and the beautification of the functional with stitched motifs drawn from the surrounding natural world. The development of sashiko was not due to one person; this woman stands for many.

Trade within Japan, between the growing cities and farming regions by way of the Sea of Japan, saw rags from the cities travel north to be used as padding and patches for garments, while farm produce filled the returning ships. The layered together rags are called *boro*. Some authors refer to boro as quilting, but to avoid confusion with modern practice, be aware that there was no 'top' constructed from seamed pieces. Since only sound fabric can hold a seam, boro is sewn over back and forth with running stitches. A boro kimono is shown in Figure 1.2.

From such basic running stitches sashiko developed. Sashiko means 'little stabs', pushing the needle through layers of cloth. Over time the interaction of lines of stitching became decorative as well as functional, and sashiko was adopted as a form of embellishment by others who were banned under the sumptuary laws from wearing colours such as red, or having silk embroidery on their clothes [Hayes, 2019]. Household items, like napkins and towels, were decorated with lines and arcs in repeating geometric patterns named for plants or geographical features which they resembled. Regional styles—*shōnai sashiko, kogin[zashi] and hishizashi*—developed over the two and half centuries of the Edo period.

Figure 1.2 Japanese rag kimono (boro kimono) dating from the Meiji period 1868–1912; purchased by the National Gallery of Victoria in 2014. Photo: National Gallery of Victoria, Melbourne; used with permission.

Kogin and hishizashi feature only horizontal stitches (Figure 1.3). A less constrained modern form is *moyōsashiko* (pattern sashiko), worked on a single layer of cloth so that 'stabbing' is not necessary. Many stitches can be loaded onto a long needle held horizontally. Such patterns, which can feature curved lines, need to be transferred onto the fabric that is to be decorated. Hitomezashi means 'one-stitch sashiko', and it is worked as if on a grid, one stitch per grid space. While a skilled eye could discern such a grid in the woven fibres of cloth or a skilled hand make perfectly even stitches, we will work hitomezashi as a counted-thread form of embroidery. 'Sashiko' can refer to any or all of these forms.

Finally, internal and external pressure forced Japan to open up to the rest of the world. The Edo period was followed by the Meiji restoration, return of power to the emperor. Mathematics as practised in Europe and North America, *yosan*, was encouraged and wusan declined [Kitagawa and Revell, 2023]. The dress regulations eased. The new authorities sought foreign expertise to restore infrastructure, neglected in the later years of the shogunate. One such expert was Dutch civil engineer George Arnold Escher, father of the graphic artist M. C. Escher. In

Figure 1.3 Some rhombic kogin motifs and fabric printed with a traditional hemp leaf motif *asanoha* common in sashiko and woodwork. Stitching and photograph by the author.

Escher's childhood home, there were items that his father had brought back to the Netherlands, such as Japanese prints, lacquer-work and a wooden puzzle. Unconventional perspectives and mathematical themes pervade his art, though he was not trained formally in mathematics; the mathematical art community claims him as one of their own.

1.2 SASHIKO POST-EDO

Boro garments and the skills of sashiko stitching were in danger of being lost in the first half of the twentieth century. Hayes (2019) traces the recognition of the value of sashiko by the exponent of the *mingei* (folk art) movement Yanagi Sōetsu (1889–1961), to a revival of its teaching, not mother to daughter but in studios and interest groups. Mingei is characterised as being hand-made by anonymous makers, functional, inexpensive, used by the masses and having regional characteristics; it exemplifies the beauty of the everyday. Exhibitions of Japanese domestic textiles travelled to galleries and museums in Europe and the USA [Shaver, 1992, Vincentelli, 2011].

Member of the Japanese avant-garde Chiyu Uemae (1920–2018) produced 176 *nui* (sewn) artworks between 1975 and 1997. Apart from their scale, these works comprised of vertical lines of close-packed stitches, executed by a male artist and purchased by collectors and galleries, echo the humble boro zokin. Self-taught, he acknowledges the influence of his background working in a textile factory, and the affordability of thread compared to his previous medium of oil paint [Uemae, 1997]. Nevertheless, he extols stitching in these words:

> "The art of stitching ... dates back to ancient times... However, the embellishment of clothing has generally been considered more as a folkcraft. I feel that both oil paint and textiles are materials of equal merit, and with this belief I decided to create artworks using stitching as a pure art."
>
> Chiyu Uemae
> [Uemae, 1997]

Better known for the technologically-advanced pleated materials with which he began to work in the 1980s, Japanese fashion designer Issey Miyake (1938–2022) has featured sashiko elements in garments since his earliest collections of the seventies.

The northern part of Honshū was profoundly affected in 2011 by the Great East Japan Earthquake, the associated tsunami and the subsequent nuclear power plant accident. In the midst of this destruction, the Ōtsuchi Reconstruction Sashiko Project was born, to provide a constructive, healing activity, particularly for women in the evacuation shelters, who had lost the structure of their daily lives, but could not help with the heavy manual labour of the massive clean-up. Hand-sewing required relatively few supplies, and no infrastructure. In time, as the items sold, grandmothers and women whose workplaces had been destroyed found an income stream. Ten years on, the enterprise became self-sufficient and the new name Ōtsuchi Sashiko was adopted. See the account of Futatsuya (2018) for a fuller description, with pictures.

Samurai Blue, the Japanese men's football team, ran out for the World Cup in 2018 wearing jerseys of deep indigo blue, decorated with vertical lines of faux white stitching. The inspiration for the design was traditional sashiko, the stitches binding together the hopes of players, supporters and staff. This choice elevated sashiko alongside previous inspirations for the uniforms: Mount Fuji, the tempering marks on steel swords, and swirling flame (which features in depictions of the deity Fudō Myōō).

Figure 1.4 Hitomezashi used for mending some trousers. Stitching and photograph by the author.

An influential book in bringing sashiko to the needles of readers of English is Susan Briscoe's *The Ultimate Sashiko Sourcebook* (2004). The visible mending and mindfulness movements have embraced sashiko, with its underlying sustainability principles and the meditative experience of slowly stitching in a deliberate way. The voices of Harumi Horiuchi and Saki Iiduka speak authentically on these themes from a Japanese perspective; I don't presume as a non-Japanese person to summarise their thoughts. There has been something of an explosion in the last five years in the number of books on sashiko available in English, both translated from Japanese (like *Simply Sashiko* [Nihon Vogue, 2020]) and written in English (such as *Make and Mend* [Marquez, 2018]). I repaired the knees of my daughter's trousers, shown in Figure 1.4, using hitomezashi. Contemporary sashiko designers add to the repertoire of sashiko patterns; AYUFISH int. is a former space engineer whose hitomezashi motifs draw on the modern built environment of skyscrapers, bridges and tetrapods.

In this book, however, we explore hitomezashi from what seems to be a novel perspective, as a form of mathematical art. Other Japanese crafts to have been considered from such a point of view are origami (Figure 1.5) and *temari* (Figure 1.6). Temari were originally made as toys, the core and the decoration being silk from worn-out kimonos.

1.3 HITOMEZASHI PRACTICALITIES

To treat hitomezashi mathematically, a very precise definition of the stitching that we will—or will not—include is required. As attractive

Figure 1.5 Origami. Initially ceremonial, paper-folding could be a recreation when paper became more accessible in the Edo period. Folding and photograph by the author.

Figure 1.6 Temari features geometric designs embroidered on a sphere. Photograph by the author.

as it can be, we exclude one-stitch sashiko which incorporates lines of diagonal stitches, or half stitches, or stitches which cross, or a form with threads woven through the stitching, *kugurizashi.*

In this book, when you come across a mathematical statement about hitomezashi, this should be taken to refer only to running stitches worked in two perpendicular directions on the square grid, with equal stitch length on the front and the back of the work. The stitching is fully packed, worked along every horizontal and vertical grid line. If this requirement is being deviated from, this will be stated explicitly. In common with other forms of sashiko, hitomezashi is traditionally worked beginning at an edge (not at the centre like, say, cross-stitch). All lines of stitching in one direction are done first, and then all lines in the perpendicular direction are completed.

In order to make the hitomezashi pieces in this book, which hereafter are referred to as samplers, you will need:

- fabric to stitch on, such as aida cloth, even-weave linen or hessian;
- thread;
- a needle (the longer the better);
- two pairs of scissors: one for fabric, one to snip thread;
- a ruler or tape measure, marked in inches and centimetres.

If you would like to draw the designs, rather than stitching them, you will need grid or graph paper, and a well-stocked pencil case. Drawing still requires one to be present and intentional in the moment.

The measuring tool is required for cutting the fabric to a suitable size. This size is determined by the number of stitches to be worked and how long they will be, which in turn is determined by a property of the fabric called its thread count.

A common thread count for aida cloth is 14 count, that is, 14 holes per inch. Stitches that enter in one hole, skip two holes and come up in the third, are a good length, giving 14 stitches in three inches. On 16 count aida cloth, I prefer to make stitches of length four, giving four stitches per inch. Even-weave linen has higher count (25, 28 or even 30) and stitches are worked over more warp and weft threads. The thread count of hessian, such as that in Figure 1.7, tends to vary from piece to piece and between warp and weft. The texture of hessian, being woven

Figure 1.7 Hessian, also called burlap, makes an economical and authentic fabric for hitomezashi. Photograph by the author.

from jute, is similar to the hemp fabric which sashiko originally softened and strengthened.

For these kinds of fabric, with gaps to pass the needle through, a round-ended (tapestry) needle works well. The lack of a point also means that you are less likely to split the thread of the stitches already in place as you slide in the later stitches. For printed fabric with a close weave (such as that used in Chapter 6 for the piece *You must be friez-ing in that*), a sharp needle is required. Genuine sashiko needles are long; compare the upper needle in Figure 1.3 to the tapestry needle below it.

The thread used in the items photographed in this book includes stranded embroidery cotton (two or three strands), four-ply crochet cotton, and sashiko thread. Sashiko thread is recommended for items that will actually be worn, and for mending; it is strong and smooth and comes in many colours, not only traditional white or blue. (In this book, only *Almost Infinite Persimmon Flower* has been made with Japanese sashiko thread.) Because crochet cotton is not intended to be passed through fabric multiple times, as happens when stitching, its surface may become fluffy. Working with shorter lengths will limit this effect.

I like to use repurposed materials when I can. I have found that an unused cross-stitch kit bought at a charity shop—which should contain a piece of aida cloth, some lengths of coloured cotton threads, and a needle—provides a suitable set of starter materials for a beginner, without too much capital outlay. Such a kit may also contain a wooden embroidery hoop; hoops are not used when stitching sashiko. A distinctive palm thimble is common for sashiko, but I don't personally use one. Another tip to keep costs reasonable is to avoid custom framing by choosing a standard-sized frame in advance, planning the stitching to fit it nicely. If you become addicted to hitomezashi, you can begin to invest in sashiko needles, thread and fabric.

Some, but not all, of the samplers in the following chapters are shown made into a finished item of some kind, such as a greeting card, a wall hanging or a brooch. Other finishing ideas are found in most sashiko books; *Sashiko* [Clay, 2019] has a particularly nice variety. Such finishing is not crucial to appreciating the mathematics discussed and is entirely optional. The first sampler appears here at the end of this chapter. In later chapters, stitching tasks will be used to introduce the mathematical idea of the chapter.

1.4 SAMPLER: THE GREAT SINE WAVE

You will begin as many a novice stitcher has around the world and throughout time: by copying a piece that is in front of you in Figure 1.8. The shifts in colour are due to the use of variegated thread to add interest to the simple pattern.

Vertical stitched lines on their own form a traditional hitomezashi pattern *kawari tategushi*, where tategushi means vertical lines and kawari means variant or changed. The variation here is that the stitches in the lines are neither all aligned nor strictly alternating. The name for horizontal lines of stitching is *yokogushi.*

Required materials:

- Fabric, thread, needle, and scissors; or grid paper, pencil, eraser and ruler.

Purpose: This sampler is intended to develop the skills of working at a good tension (without loose stitches or puckered fabric) and of 'turning' from one line of stitching to the next. Stitching is worked continuously in a manner described as boustrophedon (as the ox turns), and not in the way we write.

Instructions (stitching): Thread your needle with a length of thread that is generous enough that you won't have to join it too often, but not so long that it tangles. Knot the thread so you can anchor it. If you need to join on more thread as you go, it is fine to knot it on the reverse of the work to the thread that is coming to an end.

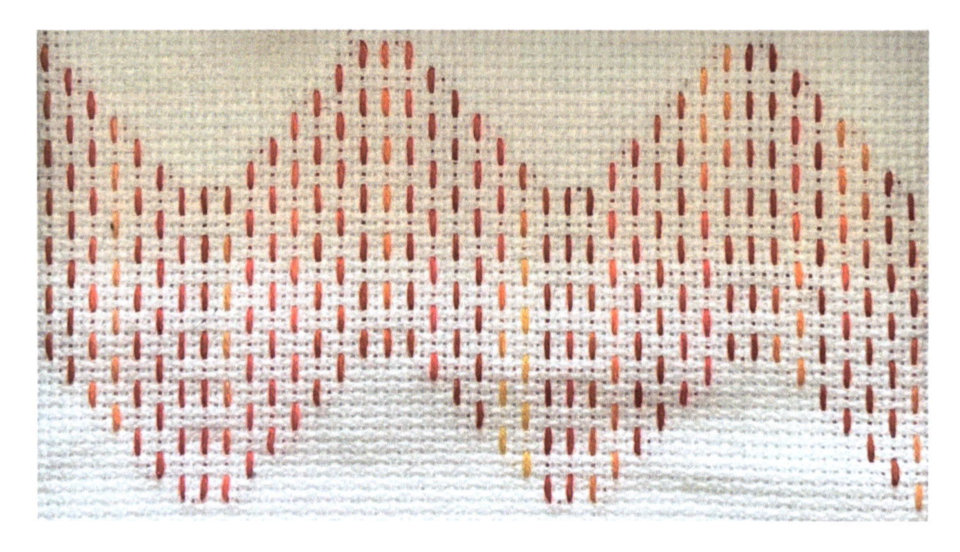

Figure 1.8 Copy this model to practice the correct tension and turning technique between lines of stitching. Stitching and photograph by the author.

Work a vertical line of running stitches, with seven stitches on the 'right' side or front of the fabric (alternating with six on the back). Your needle should pass through the gaps between the threads of your fabric, be carried over the number of warp/weft threads or holes that you have chosen, and then pass to the opposite side of the fabric through the correct gap. If you find stitching vertically to be difficult, rotate your work while you do it.

The spacing between the lines must be the same as your stitch length. To begin the second line of stitches, make a diagonal stitch on the back (wrong side) and stitch a line of running stitches of the same length, the stitches in this second line aligning with the gaps in the first. Repeat until you have seven such lines completed. Together they will look like a parallelogram. The next two lines of stitching are worked to be exactly aligned with line seven. The stitches on the back that are used to change from line seven to line eight, and from line eight to line nine, will be horizontal, not diagonal.

Continue to work lines of stitching as shown in Figure 1.8 until you have completed several cycles of the wave pattern. I had chosen a frame for my piece, shown in Figure 1.9, which determined how many lines I stitched. This frame originally held a postcard, and your piece will be about the right size to make into a greeting card if you wish.

Instructions (drawing): Replace each line of running stitches described above with a line of dashes which match exactly to the grid

Figure 1.9 Named in homage to Hokusai and to mathematics: *The Great Sine Wave*. Stitching and photograph by the author.

of your paper. That is, draw one section of grid, skip one section of grid, and so on. If you find drawing vertically to be difficult, rotate your work while you do it.

1.5 MORE TO EXPLORE

In most chapters, this section will suggest delving further into the mathematics or making more hitomezashi. In this introductory chapter, it provides tips for getting ready to stitch.

- If you need to, learn to work running stitch either by asking a friend, or from an online video tutorial.
- Locate the fabric you intend to use for stitching, and determine the number of stitches per inch (or per 5 centimetres) that this fabric can support.
- A knot worth learning for tightly joining threads when one has a short end is the weaver's knot *hatamusubi*.

CHAPTER 2

Counting: 0, 1, 2

HITOMEZASHI pattern instructions are generally given as charts, but in this chapter we will establish a different way of specifying them, using strings of binary digits. In a small sampler we will observe some of the mathematical features of hitomezashi that will be explored in more detail in the subsequent chapters. Although we are working hitomezashi as a form of counted thread embroidery we will find that many of its features can be classified simply by counting 0, 1, 2.

2.1 SAMPLER: STITCH IN LINE (PART ONE)

Required materials:

- Fabric, thread, needle and scissors; or grid paper, pencil, eraser and ruler.
- Optional: coloured pencils.

Planning: This hitomezashi sampler will (in the next step) have 36 horizontal lines of stitching. You can decide how wide you would like to make your piece; that is, how many vertical lines of stitching to use. The overall pattern repeats after six vertical lines of stitching, so a multiple of six will give a neat effect. These vertical lines should consist of 35 stitches (counting both the stitches on the front and on the back of the fabric).

Instructions: Starting at one side of the area you intend to stitch, work all the vertical lines of stitching following the pattern shown as a chart in Figure 2.1, repeating it as desired. In this chart, the underlying grid is grey, and the stitches on the front of the fabric are shown as red line segments.

DOI: 10.1201/9781003392354-2

Figure 2.1 The red line segments in this chart represent the stitches worked on the front of the fabric in the vertical lines of the sampler *Stitch in Line*. Created by the author.

Reflection: Did you find yourself thinking 'start on the front'/'start on the back', or even 'on'/'off' as you came to start a new line of stitches in your sampler?

2.2 BINARY STATES

Running stitch lines can be formed in one of exactly two ways. We will label these two states using the digits 0 and 1, with 1 corresponding to the first stitch in a line being present on the front, and 0 corresponding to it being absent. That is, 0 corresponds to the first stitch in a line being worked on the back.

Throughout this book, when specifying a hitomezashi pattern, '1' corresponds to the first stitch in a line being present on the front of the work, and '0' corresponds to it being absent on the front.

Having chosen how to start a line of running stitch, and how long to make it, there are no further choices to make. If the first stitch in the line is on the front, indicated by 1, then so too will be the third, fifth, seventh and so on. In between these, the second stitch, the fourth, the sixth and so on will be on the back. And, *vice versa*, if the first stitch in the line is on the back, indicated by 0, then so too will be the third, fifth, seventh and so on. And in between these, the second stitch, the fourth, the sixth and so on will be on the front. Using an odd number of stitches in each line gives the nice feature that the first and last stitches are either both on the front or both on the back, regardless of which end you start stitching from. Specifying that the first stitch in a line is present (or absent) on the front determines the state of all the other stitches in that line. This is the quintessential feature of running stitch.

The two digits 0 and 1 are the building blocks of the *binary number system* which operates in base two, rather than the familiar base ten system we all use everyday. It's the number system used by computers. An ordered list made up of the digits 0 and 1 interpreted as a number in base two tells us which powers of two are present or absent in that number. For example, the base ten number 13 (meaning $1 \times 10^1 + 3 \times 10^0$) is written in base two as 1101_2 meaning $1 \times 2^3 + 1 \times 2^2 + 0 \times 2^1 + 1 \times 2^0$. The two digits 0 and 1 are also used in two-element *Boolean algebra*, where they represent true (1) and false (0). In digital circuits, the operations of Boolean algebra and calculations in base two can be carried out utilising switches in 'on' and 'off' states.

We are using 0 and 1 to specify two states—we are not going to perform any calculations or logical operations with them! We could have used two different symbols, or assigned the opposite meanings to them. However, our labelling is both obvious and easy to remember and follows the usage established in the *Math Art Challenge* [Perkins and Seaton, 2020].

The stitching pattern shown in Figure 2.1 for the vertical lines in our sampler *Stitch in Line* can be written using this labelling as a *binary string*

$$101101\ldots101$$

The ellipsis (...) indicates an unspecified number of repeats, following the established pattern. To implement this as a stitching instruction, another piece of information is needed: how many stitches to work in each line. Nevertheless, compare how much more compact this is than the chart in Figure 2.1.

0	1	0	1	1	0	1	0	0	1	1	0	0	0	1	1	0	1	0	1	0	1	0	0	1	0	1	0	1	1	0	0	0	1	0	1

Figure 2.2 The binary string to be followed to complete the horizontal lines for the *Stitch in Line*. Check boxes are supplied to keep track of progress as the lines are completed. Created by the author.

We will explore a large variety of ways to choose the binary strings for hitomezashi in subsequent chapters. But for now, with the labelling system established, we can come back to our sampler to add in the horizontal lines of stitching.

2.3 SAMPLER: STITCH IN LINE (PART TWO)

This is where the magic happens! The binary string we will use as the instructions to add the horizontal lines of stitching is given in Figure 2.2.

It's more convenient (in a book) to write out the stitching instructions across the page, but remember that these are for the horizontal lines of stitching. The first digit 0 specifies the state of the bottom line of stitching, the second specifies the line above it, working from the bottom to the top, with the last row of stitching to be formed, the top one, being in state 1.

Figure 2.3 shows the sequence of horizontal lines worked on their own, and the vertical lines also worked on their own. The sequence for the horizontal lines is not obvious or easy to remember, so ticking the lines off as you go, in the boxes provided in Figure 2.2, will help you avoid errors or let you put down your work and pick it up again. (If you find this helpful, please note you will need to create such an aid yourself for the remaining samplers in the book.)

Reflection: Once you have finished your stitching or drawing, take a bit of time to look at your sampler. What do you notice? What do you wonder? If you haven't finished yet, but you want to keep reading, Figure 2.4 shows the completed stitching.

2.4 VERTEX DEGREE AND MAP COLOURING

In looking at the sampler, you might have noticed some symmetry or, if you stitched it, wondered about the interesting complementary pattern formed on the back of the work. Perhaps the most striking thing—the magic so to speak—is that, as the horizontal lines fall into place, the

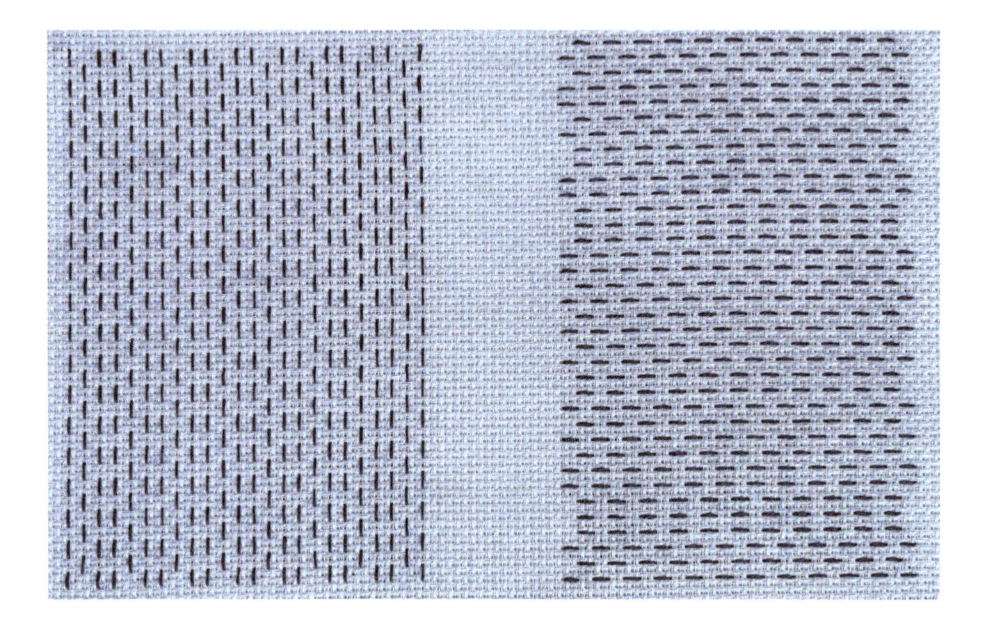

Figure 2.3 The stitched horizontal stitching lines defined in Figure 2.2 on the right. On the left, the vertical stitched lines are shown. Stitching and photograph by the author.

stitches interact and link with the individual vertical stitches. The connected stitches form stepped lines from one side of the piece to another, and polygons, some of these enclosed within larger polygons. The lines and polygons outline regions in our design.

A *polygon* is a closed figure lying in the plane which has a particular number of *vertices* (points) joined by an equal number of *edges* (line segments), although the actual meaning of the word is 'many angles'. Apart from the small squares that form, the other polygons we can see are *concave*. For an intuitive feeling for what it means for a polygon to be concave, compare the crosses in the sampler, which have twelve edges, with the Australian fifty-cent piece in Figure 5.2 in Chapter 5, which is a regular, *convex* dodecagon. The shape outlined by each hitomezashi polygon is a *polyomino*, formed from identical squares joined edge-to-edge.

We will often refer to the polygons as *loops*. Here we are thinking as topologists do. *Topology* is the mathematical study of connectivity. Two objects can be thought of as the same in a topological sense if they can

Figure 2.4 The finished hitomezashi sampler *Stitch in Line*. Stitching and photograph by the author.

Figure 2.5 The regions outlined within *Stitch in Line* can be coloured using only two colours. Drawing and photograph by the author.

be stretched or twisted, but not cut or glued, to resemble each other; in topology, loops can have corners.

If you draw your sampler, you might be tempted to start colouring it in. Children particularly seem to like doing this, as soon as the loops begin to close. You may be tempted to use lots of colours, but in Figure 2.5 only two colours are used. Yet each region is differently coloured from its neighbours, the regions with which it shares any edges. The hitomezashi divides the surface of our sampler into regions, as the borders on a geographical map do for countries. The *chromatic number* of a map is the smallest number of colours needed to be used so that two adjacent regions (countries) have a different colour when the map is coloured in. We might wonder whether there is something special about our sampler, or ask if all hitomezashi designs have chromatic number two.

We will use a simple example of a *proof by contradiction* to establish that our sampler is not special in this regard. We will ask what would

Figure 2.6 If each brick in the wall on the left was to be coloured differently from any with which it shares an edge, three colours would be required. This is illustrated on the right using a small piece of hexagonal patchwork in progress. Stitching and photographs by the author.

follow logically if a hitomezashi design did have chromatic number three, and see that it leads us to something we know to be false.

We have stitched our hitomezashi designs on a *square grid*; the grid lines meet at vertices. The *degree* of a vertex in a grid is the number of grid edges that join it. For a square grid, every internal vertex has degree four. But we can also ignore the grid and think only about how the vertices are connected by the stitches on the front of the hitomezashi design. Because our hitomezashi is formed from vertical and horizontal lines of alternating stitches, each vertex has degree exactly two when considered as part of the design; one horizontal and one vertical stitch connect to it. The vertices around the edges of a finite piece of hitomezashi may have degree lower than this, but not higher.

If our map were to have chromatic number higher than two, there would have to be somewhere within it a vertex of degree three, like the ones we encounter if we try to colour in some standard brickwork or piece together patchwork made of hexagons (see Figure 2.6). There's an old joke that a topologist is someone who can't tell a doughnut from a coffee cup; here, we don't distinguish between a brick wall and a bedspread! Since our hitomezashi patterns don't ever have such vertices, they all have chromatic number less than three. But we also have a *counterexample* (Figure 2.5) for someone who might try to argue that it is one. Hence it must be two in general.

2.5 MORE TO EXPLORE

- If you drew your sampler *Stitch in Line*, but didn't colour it in, you could do that now.
- If you stitched your sampler *Stitch in Line*, now draw it, and colour it in.
- If you drew your sampler *Stitch in Line*, stitching it—even with thread on card if you have no suitable fabric—will be very informative for the next chapter.
- To find out more about chromatic numbers and map colouring, look up the Four Colo(u)r Theorem.

CHAPTER 3

Loops

HITOMEZASHI stitching lines interact to form long paths and closed polygons or loops, with constraints on their structure. Properties of these loops involve modular or clock arithmetic. In this chapter, we discuss some general properties of hitomezashi loops and investigate the properties of the traditional persimmon motif.

3.1 SAMPLER: ALMOST INFINITE PERSIMMON FLOWER

Required materials:

- Fabric, thread, needle and scissors; or grid paper, pencil, eraser and ruler.

Planning: The design we will stitch is traditional, called *mugen kakino-hanazashi*, infinite persimmon flower stitch. This design is most naturally worked on a square piece of fabric. It is probably simplest to stitch this piece beginning with the central lines of stitching, rather than at the edge as is customary, as shown in Figure 3.1.
Instructions: The stitching instructions for both directions can be written as:

$$\ldots 0101010\mathbf{11}0101010\ldots$$

Two aligned vertical lines of stitching interact with two aligned horizontal lines of stitching in such a way that a single small square is formed exactly in the centre of the design. Apart from these four particular lines of stitching, the states of the lines of stitching alternate from one to the next, to the edges of the area to be worked.
Reflection: What do you notice about the geometric shapes formed on this sampler? What is on the reverse? What do you wonder?

DOI: 10.1201/9781003392354-3

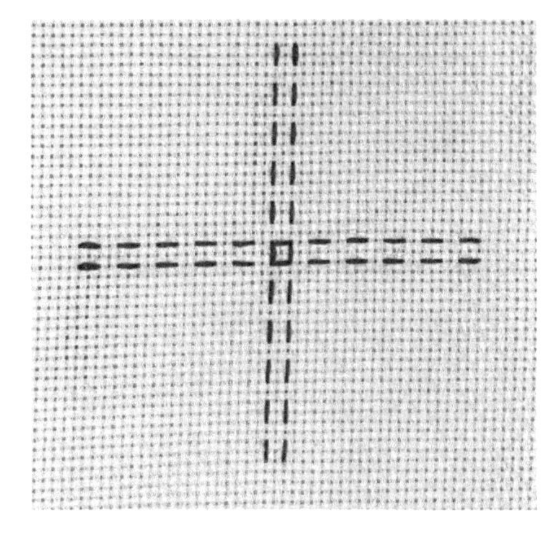

Figure 3.1 Starting the *Almost Infinite Persimmon Flower* sampler working from the centre, with two aligned lines of stitching in each direction. Stitching and photograph by the author.

3.2 PERSIMMON PROPERTIES

The traditional design we have stitched, seen in Figure 3.2, consists of nested polygons each of a structure which we will refer to as a *persimmon*. The design is highly symmetric and it can be extended, if not literally to infinity, to the edge of our fabric.

Kakinohanazashi literally means stitch (zashi) of the flower (hana) of the persimmon (kaki). The photograph in Figure 3.3 shows a very ripe persimmon fruit—they are highly astringent and not enjoyable unless extremely soft. The small creamy white flower which precedes the fruit has four petals, and the four sepals remain on the golden fruit. The single and double persimmon stitched on the coaster mimic the sepals. Traders introduced persimmons to Europe from Japan and China in the nineteenth century.

The first two polygons, drawn in Figure 3.4, enclose polyominoes consisting of 1 and 13 grid squares, respectively. We call this the *area* of the polyomino (and casually the area of the polygon). The number of stitches which link together to form these polygons are 4 and 20, respectively. We call this their *perimeter length*.

A traditional design consisting solely of repetitions of the small squares is called *kuchizashi* meaning 'mouth stitch'. A traditional design

Figure 3.2 *Almost Infinite Persimmon Flower.* Stitching and photograph by the author.

Figure 3.3 A ripe persimmon fruit and a coaster featuring kakinohanazashi. Stitching and photograph by the author.

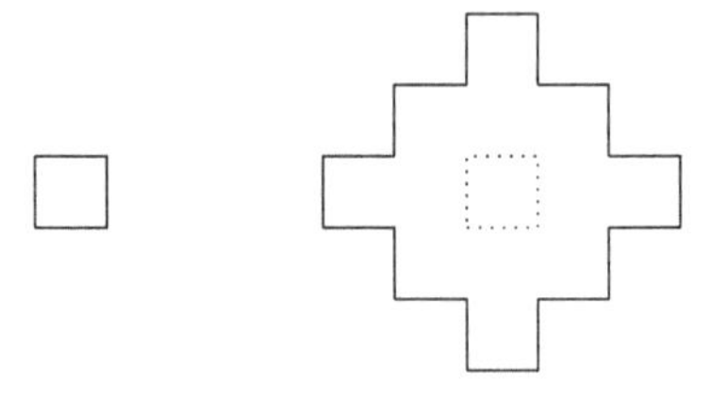

Figure 3.4 Persimmon polyominoes of height 1 and 5. Created by the author.

consisting solely of repetitions of the height five persimmons is kakino-hanazashi (without the descriptor mugen meaning infinite).

On the reverse of the fabric is a very similar design of nested persimmons, but radiating out from a cross. The area enclosed by the cross (see Figure 3.5) is 5, and its perimeter length is 12. A traditional design consisting solely of repetitions of the cross is called *jūjizashi* meaning 'ten-cross stitch'. The character for the number ten resembles a cross.

Imagine drawing the smallest rectangular box which encloses a polygon. We will call the height of the box the *height* of the polygon. Each of the nested persimmons fits into a square box, so its height is the same as its *width*, the width of this box. The single square has height 1, and the cross has height 3. Alternating between the back and front of the fabric, each persimmon is taller and wider by 2 squares than its predecessor, so the heights are successive odd numbers. Count the squares and stitches to complete Table 3.1. How does symmetry reduce the amount of counting needed? Are patterns emerging in the numbers? If you think you can see a pattern, predict the area or length of a polygon for which you haven't done the counting yet. Then count the length in stitches and the area in grid squares to test your conjecture.

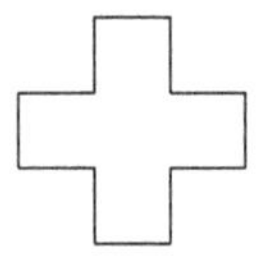

Figure 3.5 From the reverse of the sampler, a ten cross, but in this context we regard it as a persimmon polyomino of height 3. Created by the author.

TABLE 3.1 Perimeter length and area of the persimmon polyominoes

Height	Length	Area
1	4	1
3	12	5
5	20	13
7		
9		
11		

3.3 PARITY AND CLOCK ARITHMETIC

The parity of a number expresses whether it is odd or even. As a mathematician, I was struck immediately by the two- and four-fold symmetries of hitomezashi. My colleague Carol Hayes drew my attention to all the odd numbers that she noticed in kakinohanazashi. Not only are the heights and widths odd numbers, but each row or column forming the polyominoes, each cross-section, also contains an odd number of squares. A scholar of Japanese culture and language, she knew that odd numbers are considered more favourably in Japanese culture than even numbers, and that arrangements consisting of odd numbers of objects contribute to the aesthetic experience of *fukinsei*, usually translated as asymmetry but also carrying a sense of incompleteness, as the monk Kenkō wrote:

> "In everything, no matter what it may be, uniformity is undesirable. Leaving something incomplete makes it interesting, and gives one the feeling that there is room for growth."
>
> Kenkō (c. 1330)
> From Essay 82, translated by Donald Keene [Keene, 1967, p.70]

The famous *Ryōan-ji* dry landscape garden is an example of an auspicious numbers garden, containing fifteen rocks arranged in three groups consisting of three, five and seven rocks respectively [Deane, 2012]. In *ikebana* the asymmetry achievable when arranging three flower stems gives a dynamic feeling, a sense of movement. Carol Hayes both wrote and translated *tanka* poems, which consist of thirty-one syllables arranged as seven-five-seven-seven-seven. *Shichi-Go-San* (seven-five-three) is a festival celebrating healthy growth to the age of three and seven for

Figure 3.6 Two sets of coasters, one a gift from Europe, the other a gift from Japan. Photograph by the author.

girls, and five for boys. 'Four', on the other hand, is a homophone for 'death'. The cutlery in my Australian home was bought as a set of eight, my plates as a set of four, and the coasters shown in Figure 3.6 acquired as a set of six. In Japan, five is more usual for homewares such as sake cups, rice bowls, chopsticks and indeed coasters.

The 'oddness' of an odd number of objects consists of there being one left over, literally the 'odd one out', when they are lined up in pairs. *Modular arithmetic* generalises the idea of parity (paired-ness). Even numbers are congruent to 0 modulo 2, and odd numbers are congruent to 1 modulo 2. More generally, if a and b have the same remainder when divided by k (where a, b and $k > 0$ are integers) we write

$$a \equiv b \bmod k$$

where the tribar symbol expresses *congruence* modulo k.

We use arithmetic modulo 60 and modulo 12 (or modulo 24) when we talk about time, the reason it is also called *clock arithmetic*. Suppose I start stitching at 50 minutes past 11 in the morning. If I stitch for 90 minutes, the time when I stop is not called '140 minutes past 11', or even '80 minutes past 12'. Using a 12-hour clock, we say it is 20 past 1; using a 24-hour clock, we say it is 13:20. The pattern for the perimeter lengths of the nested persimmons is fairly straightforward to spot, by comparing the numbers in the first two columns of Table 3.1.

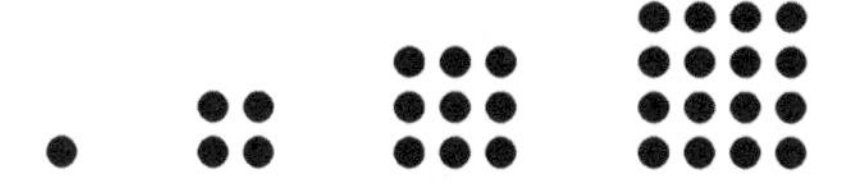

Figure 3.7 Arrays of dots showing the first few square numbers. Created by the author.

The perimeter length is a multiple of 4, reflecting the four-fold rotational symmetry of the polygons. Labelling the heights as $2n-1$ for $n=1,2,\ldots,$ the corresponding perimeter length is $4(2n-1)$. The perimeter length of each polygon $8n-4$ is congruent to 4 mod 8.

3.4 POLYGONAL NUMBERS AND PERSIMMON AREA

Polygonal numbers enumerate objects arranged in regular arrays. We will use square and triangular numbers to give several *visual proofs* for the area enclosed by the persimmons.

Square numbers count the number of dots in arrangements shown in Figure 3.7. Labelling them with $n = 1,2,3,\ldots,$ the nth square number counts the number of dots in n rows each containing n dots: n^2 in total. We read 'n to the power 2' as 'n-squared' almost unthinkingly.

The nth *triangular number*, indicated by T_n, is drawn by placing n dots in a base row, $n-1$ in the row above, one fewer again in the row above that and so on, until the final top row contains a single dot. These can be drawn as an equilateral triangle by staggering the dots between rows, but it suits our purposes to draw them using a right-angled triangle as is done in Figure 3.8. The values are 1, 3, 6, 10, ... and in general

$$T_n = 1+2+3+\ldots+n = \frac{n(n+1)}{2}.$$

To find a pattern for the areas of the persimmons, consider Figure 3.9. It shows the persimmon of height 5 ($n=3$) inside the smallest square

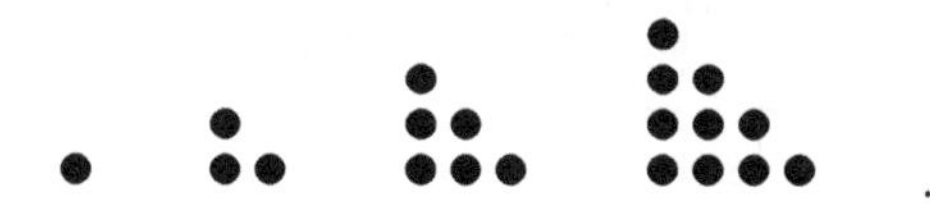

Figure 3.8 Arrays of dots showing the first few triangular numbers. Created by the author.

Figure 3.9 The persimmon of height five drawn within a square box of height five, leaving a triangle of height two at each corner. Created by the author.

box that contains it. Around the polyomino there are four 'triangles' which each contain T_2 grid squares.

In general, the area of the nth persimmon is

$$C_n = (2n-1)^2 - 4T_{n-1} = 2n^2 - 2n + 1.$$

The numbers in this sequence 1, 5, 13, 25, 41, ... are called the *centred square numbers*, and we use the symbol C_n for them.

Other relationships with the square and triangular numbers provide additional visual proofs of this rule. Let's colour the interior squares of some persimmons like the squares of a chess board (see Figure 3.10). In each one we can see two adjacent ordinary square numbers superposed, one consisting of black squares and the other of white. This gives a different grouping of the terms, but the same values for

$$C_n = n^2 + (n-1)^2.$$

It's also possible to decompose them in terms only of triangular numbers:

$$C_n = 4T_{n-1} + 1.$$

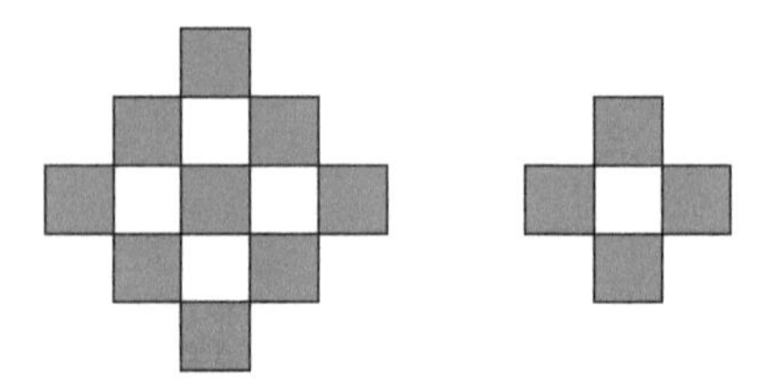

Figure 3.10 Chessboard colouring of persimmons of height three and five. Created by the author.

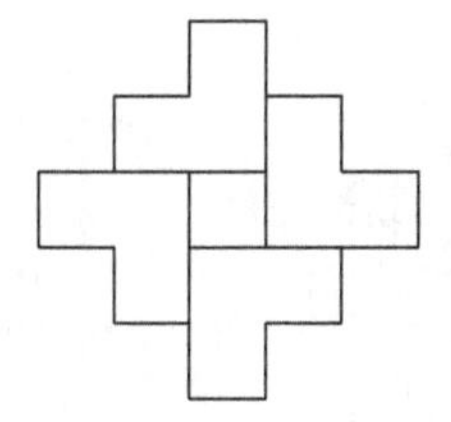

Figure 3.11 Persimmon of height five constructed from four triangular numbers and a square. Created by the author.

Figure 3.11 shows (for $n = 3$) four triangular number arrangements of squares surrounding a single central square.

It's clear that each centred square number $C_n = 2n(n-1)+1$ is an odd number, but if we look more closely, we can see that the first term has a factor of four. Because they are consecutive, one or the other of n or $n-1$ must be an even number, that is, contribute a factor of two to the product $n(n-1)$. Thus the centred square numbers, that is, the area of the hitomezashi persimmons, is congruent to 1 modulo 4.

3.5 LENGTHS AND AREAS OF HITOMEZASHI LOOPS

The polygons in our infinite persimmon sampler are highly symmetric. In Figure 3.12, some nested loops that feature in the random sampler *Ever so Airy a Thread* (coming up in Chapter 5) are shown. Determine the length, height, width and area of each of these loops. What do you notice?

The following things have been proved to be true about all fully packed hitomezashi loops:

- any horizontal or vertical cross-section contains an odd number of grid squares, and in particular the width and height are odd;
- the perimeter length is congruent to 4 modulo 8;
- the enclosed area is congruent to 1 modulo 4.

The first result is due to Pete (2008), the second and third to Defant and Kravitz (2024); a succinct proof for the second result was subsequently given by Ren and Zhang (2024). In proving these results, the *Mirror Property* of loops was established and used. Stating this as non-technically as possible, it means that sections of the boundary of a loop

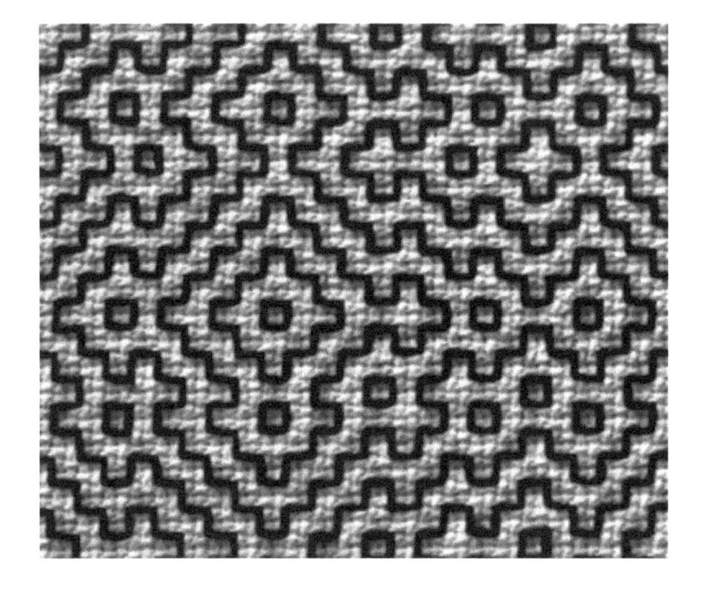

Figure 3.12 Several loops from a random piece of hitomezashi. Stitching and photograph by the author.

on the left and right must be mirror images of each other across a vertical axis, and so too for sections of the top and bottom. Coming back to the idea of drawing a box around a hitomezashi loop, below any stitch in contact with the box at the top there will be a stitch in contact with it at the bottom, and so too for left and right. In between these extremal stitches there is no other stitch which forms part of the perimeter of the same polygon. These structural properties of hitomezashi loops explain the striking symmetry; we explore this symmetry more deeply in Chapter 6.

3.6 COUNTING HITOMEZASHI LOOPS

In a regular hitomezashi pattern, the loops are of known size and spacing, and by calculation it is possible to determine exactly how many loops will appear if that design is stitched on a piece of fabric of a certain size. But is there any way of knowing how many loops will form if the state of each line of stitching is determined randomly? This was a question that many participants in the *Math Art Challenge* asked.

Of course, we could count them, for any finite design we look at. But what can we say in general, about arbitrary loops, with the state of each line of stitching decided randomly? Probability theory gives a way to calculate the *expected value* of the number of loops, an average weighted by the probabilities. Obviously, this depends on the number of horizontal and vertical lines of stitching, m and n. The expected number of loops grows with the grid size according to this rule [Defant and Kravitz, 2024]:

$$\frac{\pi^2 - 9}{12} mn \approx 0.0725\, mn.$$

This kind of result is *asymptotic*, meaning that it is a better prediction for large values of m and n. The randomly designed sampler *Ever so Airy a Thread* (Figure 5.8) has $m = n = 110$. For a grid this size, the expected value rule gives 875 loops. While 110 rows and columns is a lot of stitching, these numbers are not large by the standards of asymptotics. But we can ask if this seems reasonable. If the stitches were arranged regularly to form tiny squares (kuchizashi), in the same area there would be around 3000 loops (55×55). A mugen kakinohanazashi of height 109 squares would feature 28 nested loops.

We might want to ask other questions about hitomezashi loops such as how many there are of a certain height and width, and if we can optimise the perimeter length or area. Defant, Kravitz and Tenner (2023) considered some such questions. In counting lengths and areas, we do not need to distinguish between loops that are equivalent to each other if rotated, reflected or translated, and we ignore enclosed loops.

The first interesting perimeter length to consider is 20; there is only one possible loop with length 4 (the square), and one with length 12 (the ten-cross). As well as the height 5 persimmon (see Figure 3.4), the motifs in Figure 3.13 have length 20. They have height of 3, width of 5 and enclose an area of 9.

Among the possible loops in Figure 3.14 with a perimeter length of 28, we find two traditional motifs. Labelled as (a) is a persimmon and as (b) *igeta*, which takes its name from the frame or kerb of a well. The others do not have traditional names; the motif labelled (c) was included in *Stitch in Line* and (e) is an extended version of the motif in Figure 3.13. Their features are summarised in Table 3.2.

3.7 MORE TO EXPLORE

- Draw the persimmon with height 7 three times. Annotate these drawings to demonstrate that $C_4 = 7^2 - 4T_3 = 4^2 + 3^2 = 4T_3 + 1$.

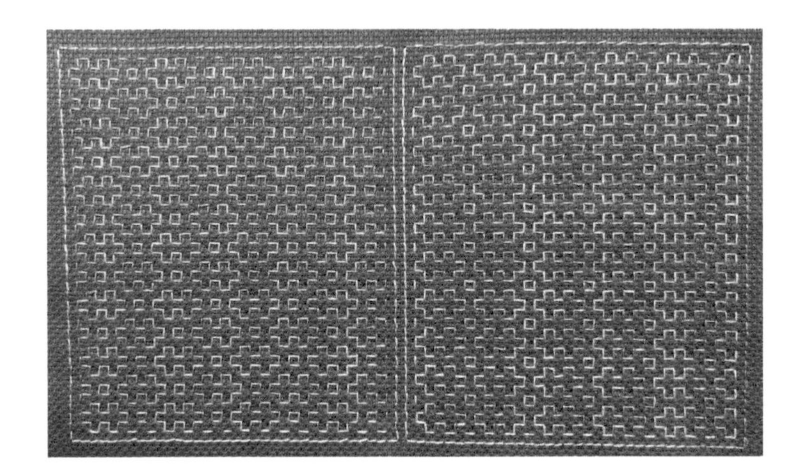

Figure 3.13 Two different regular arrangements containing loops with perimeter length 20. Stitching and photograph by the author.

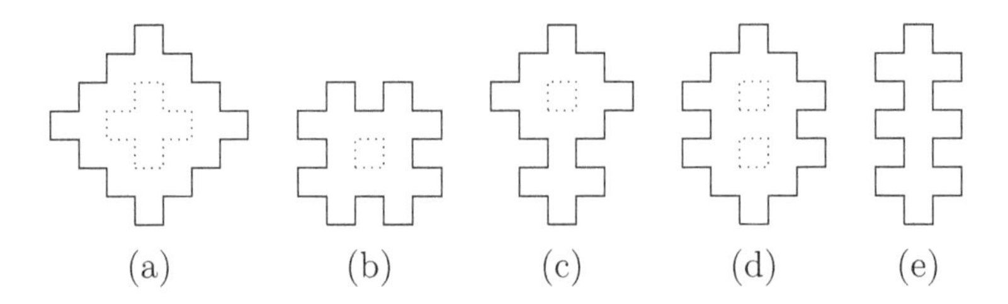

Figure 3.14 Some hitomezashi loops of perimeter length 28. Created by the author.

TABLE 3.2 Features of loops of perimeter length 28 shown in Figure 3.14

Loop	Height	Width	Length	Area
(a)	7	7	28	25
(b)	5	5	28	17
(c)	7	5	28	17
(d)	7	5	28	21
(e)	7	3	28	13

- The result $1+2+3+\ldots+n = \frac{n(n+1)}{2}$ has been stated without proof. Draw a rectangle to represent n rows containing $n+1$ dots, and colour half the dots in one colour and half in another to give a visual proof that $n(n+1) = 2T_n$.
- Square numbers, triangular numbers and centred square numbers enumerate only three possible geometric arrangements of dots or squares. Look up *figurate numbers* in general.
- Look ahead to photographs featuring random hitomezashi loops and regular loops other than persimmons. Check that these loops also have odd heights, widths and cross-sections, and that their perimeter length and enclosed area have the expected congruence, modulo 8 and 4 respectively.

II

Three Big Ideas

CHAPTER 4

On the Flip Side

DUALITY as a concept in mathematics has a deeper meaning than simply having two states. It does not necessarily have the meaning of polarity or dichotomy as it does in philosophy. Duality is pervasive across disparate areas of maths and physics, and yields powerful perspectives. Yet we can quite literally get a feel for it by making some small pieces of hitomezashi.

4.1 SAMPLER: DUALITY COASTERS

This is one of the few samplers for which drawing will not suffice to experience the subject of the chapter in a visual and tactile way. If suitable fabric is not to hand, a possible alternative would be to mark a grid onto cardboard, and then stitch with an ordinary needle and sewing thread (or string or knitting yarn).

Required materials:

- Fabric, thread, needle and scissors.

Planning: Coasters are generally about 10 cm $\times$ 10 cm in size. The examples shown in Figure 4.1 have been worked on hessian, stitching across four warp or weft threads, with a border of running stitch which is 13 stitches long in each direction. The design itself is worked on 11 stitches in each direction. Calculate how many stitches can be worked for the thread count of the fabric you intend to use; remember that having stitch lines which comprise an odd number of stitches in total (counting stitches on the front and on the back) makes working the patterns simpler.

Instructions: Given in Table 4.1 are the instructions for a set of nine coasters, specified in the form established in Chapter 2. If you do not

DOI: 10.1201/9781003392354-4

Figure 4.1 The nine *Duality Coasters*. Coasters A, B and C are brown, coasters D, E and F are dark green, and coasters G, H and I are light green. Stitching and photograph by the author.

wish to make nine coasters, make at least one set of three, such as A, B and C (or D, E and F; or G, H and I).

Reflection: Compare the fronts and the backs of your coasters. Does it matter if you turn them on a side edge (like turning the pages of this book), or on the top or bottom?

TABLE 4.1 Stitching instructions for nine *Duality Coasters*

Coaster	Vertical lines	Horizontal lines
A	111...	111...
B	111 ...	00110011 ...
C	000...	11001100...
D	1010...	111...
E	110110...	00110011...
F	001001...	11001100...
G	01100110...	01100110...
H	1010...	0101...
I	0101...	1010...

4.2 INTRODUCTION TO MATHEMATICAL DUALITY

The renowned British mathematician Professor Sir Michael Atiyah (1929–2019) gave the following non-technical description of duality in a talk that he gave in 2007:

> "Duality in mathematics is not a theorem, but a 'principle'... going back hundreds of years. ... Fundamentally, duality gives two different points of view of looking at the same object. There are many things that have two different points of view and in principle they are all dualities."
>
> Sir Michael Atiyah
> [Atiyah, 2007, p. 1]

A famous duality in physics is the *wave-particle duality*, whereby some properties of quantum objects, such as light, are best captured by considering the objects as waves, while other properties are particle-like.

A simple example of duality can be illustrated using the *Venn diagram* of a single set inside a universal set Ω, as in Figure 4.2. If we shade in A, then the unshaded region is the *complement* of A, written as A' or A^c or $\overline{A}$. This is shown in the image on the left of the figure. Alternatively, we could shade in A', and effectively we would shade *out* A, as has been done on the right. Defining or describing A effectively also defines or describes A'. These are Atiyah's two points of view.

This simple example exemplifies the property of (most) dualities, that the dual of the dual is the original object: $(A')' = A$. However, it

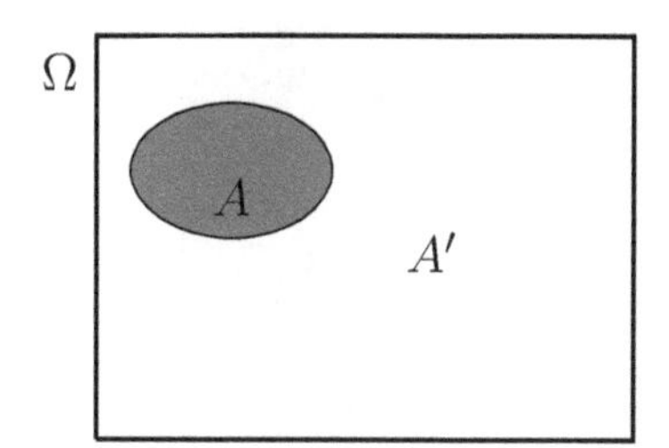

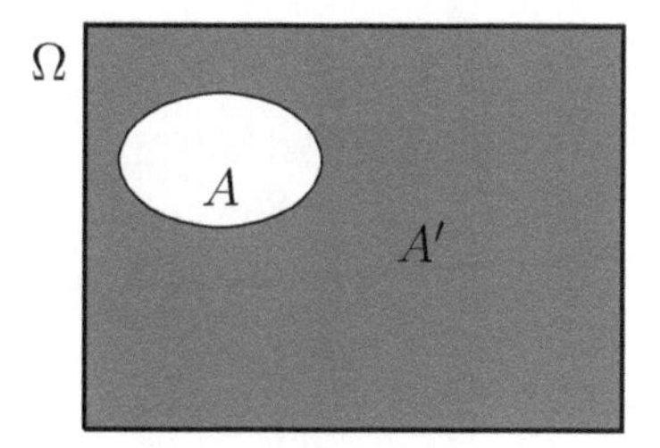

Figure 4.2 In the left-hand image, the set A has been shaded in, and its complement in the universal set is unshaded. On the right, the complement A' has been shaded in, effectively shading out A. Created by the author.

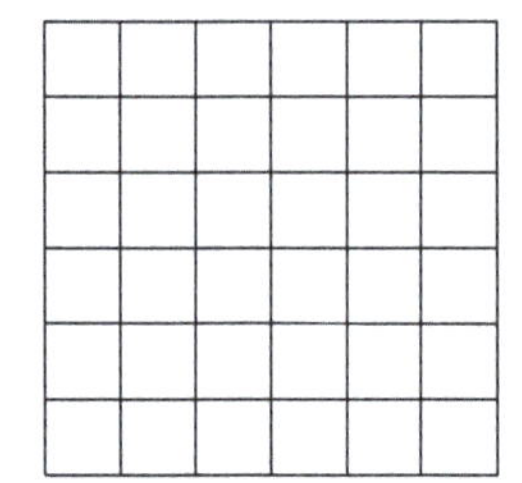
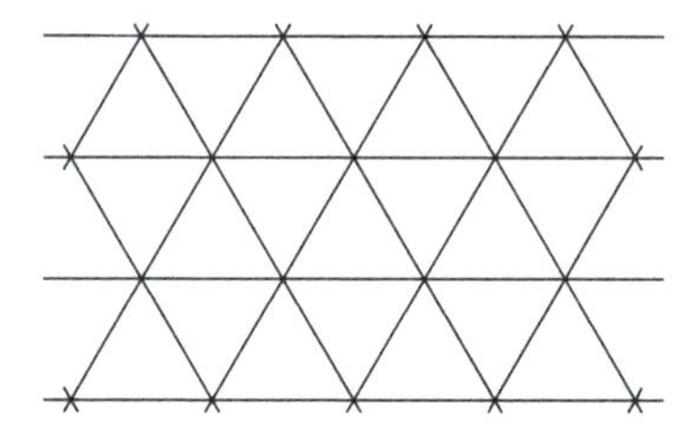

Figure 4.3 The image on the left is a square grid and the image on the right is a regular triangular or isometric grid. Created by the author.

does not illustrate the concept that an object may be *self-dual*. For an example of this latter concept we look to another duality, that of grids.

Figure 4.3 indicates two of the three possible regular grids on the plane (those in which all edge lengths are equal and all angles between edges are equal): the square grid and the isometric (triangular) grid. They each have vertices, edges which connect the vertices, and faces which are marked out by the edges; the tiles in Figure 4.4 are the faces of the third kind of grid, hexagonal. (There is an area of abstract algebra called *lattice theory* in which duality appropriately defined plays a key role. To avoid any possible confusion, we refer to grids here, although mathematicians and physicists would often say 'lattices' for these objects.)

To form the dual, place a vertex at the centre of each face of the original grid. For adjacent faces on the original grid (those which share

Figure 4.4 The grout between these tiles marks out the edges of a hexagonal grid, and the tiles are the faces. Photograph by the author.

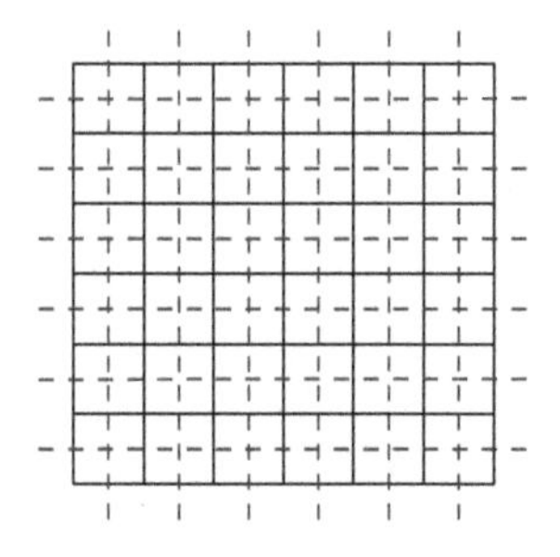

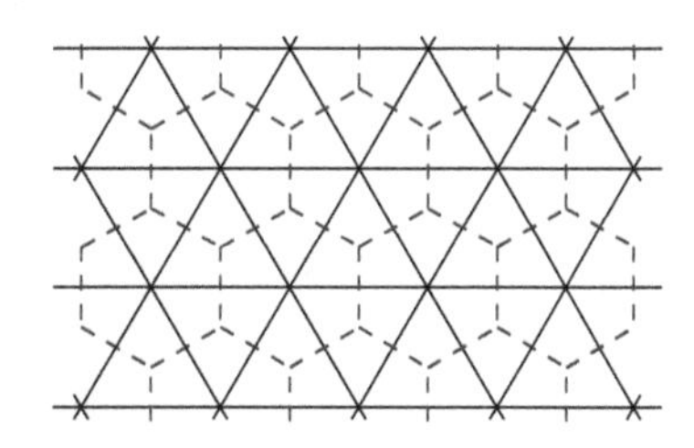

Figure 4.5 The dual of the square grid and of the isometric grid are marked in blue. The edges of the dual meet at vertices lying at the centre of the faces of the original grid. Created by the author.

an edge), connect the corresponding dual vertices with an edge of the dual grid. Superposing the duals on the originals, as in Figure 4.5, we see that the edges match up one-to-one between a grid and its dual, and the vertices of one correspond to the faces of the other.

We also notice from the image on the right that the triangular and hexagonal grids are dual to each other. On the left, we see that the dual of a square grid is a square grid. That is, the square grid is self-dual.

We can extend this kind of duality to irregularly shaped 'faces', and 'edges' not necessarily drawn as line segments. In Chapter 2 we considered map colouring, assigning colours to regions marked out by hitomezashi strands. The dual of a finite map is a finite *graph*.

Yet again we encounter a word with more than one mathematical meaning. In this case, it is the word *graph*. The meaning we usually meet first in elementary school is the visual representation of data or, in high school, plotting a mathematical relationship on a set of axes, such as the graph shown in Figure 4.6. But the meaning we want to use here is quite different, that of *graph theory*. In this context, a graph consists of a collection of vertices (or nodes) and edges connecting some or all of them.

To form the dual, to each region of the original map assign a node of the dual graph. As with grids, each edge in the dual graph corresponds to an edge separating adjacent map regions, and connects the corresponding nodes. Properties of the dual graph, which may be simpler to establish, can be applied to the original map. For example, questions about the possible colouring of a map's faces can be posed as questions about the vertex colouring of its dual graph. In this context, maps are generally referred to as planar graphs, and the regions or countries are the faces of the graph. Figure 4.7 represents the graph dual to the map of Australia.

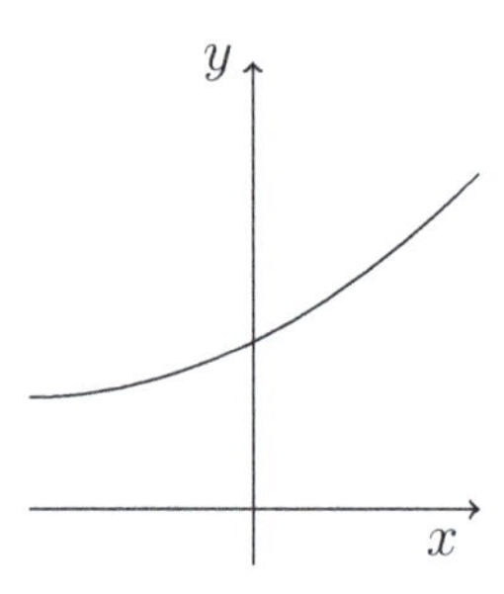

Figure 4.6 This visual representation of the function with rule $f(x) = (x+a)^2 + b$ is called the graph of f; however, it is not the kind of graph we consider in this chapter. Created by the author.

The triangles in the dual graph tell us that the map has chromatic number (at least) three. The unconnected node represents the island state of Tasmania.

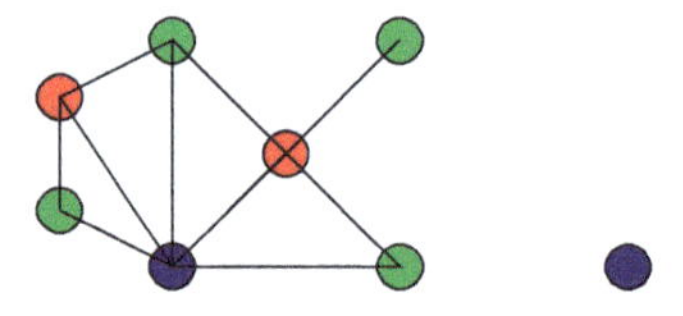

Figure 4.7 The dual graph for a map of the six states and the two self-governing internal territories of Australia. One possible three-colouring—of many—is shown. Created by the author.

4.3 DUALITY IN HITOMEZASHI PATTERNS

Not all embroidery types have the property that a complementary design of the same kind can be found on the back. In some, such as counted cross-stitch, the reverse is not intended to be seen; in such a case, the two sides may be appropriately called 'right' and 'wrong'. Both sides of some cross-stitch are shown in Figure 4.8. In other embroidery forms, such as Holbein stitch used in blackwork, the design is worked so that it is intentionally reversible; a stitch on the front is accompanied by a corresponding stitch on the back. An example is shown in Figure 4.9. Joshua Holden (2008) explored the graph theoretical aspects of blackwork in a chapter of the edited volume *Making Mathematics with Needlework* [belcastro and Yackel, 2008].

Figure 4.8 A line of cross-stitch, and below it, the reverse or 'wrong side' of a line of crosses. Stitching and photograph by the author.

Because the gap between running stitches on the front of a piece of hitomezashi corresponds to a stitch on the reverse, the designs on the two sides of the fabric are complements of each other, two patterns created by the same lines of stitching. We can think of the pattern formed on the back of a piece of hitomezashi as the dual in a way that we will make precise.

Take another look at your *Duality Coasters*, front and back. *Hana-fukin* (flower cloths), which have practical and decorative uses, showcase both sides of the sashiko they are adorned with, being stitched through two layers of cloth with knots and turning stitches concealed between the layers [Nihon Vogue, 2020]. Our coasters comprise a single layer, so as you look at the back, disregard the border and the edge effects of

Figure 4.9 A small piece of Holbein or double running stitch. Stitching and photograph by the author.

Figure 4.10 From left to right, coaster A, coaster B and coaster C. Stitching and photograph by the author.

the stitches that carry the thread from the end of one line to start the next line; focus on the central part of the coaster to see the dual pattern clearly.

Figure 4.10 shows the three coasters A, B and C. Coaster A is stitched with the traditional pattern kuchizashi (mouth stitch) which Briscoe (2022) also calls *chiisana shikaku* (small squares). This pattern is self-dual, the pattern on the reverse also being aligned squares. It might be tempting to expect that on the reverse of coaster B, which is stitched on the front with kawari kuchizashi, there would be more off-set squares. But rather, the dual is a variant of *jōkaku* (castellation). This is the pattern that has been stitched on coaster C, and naturally we find the off-set squares on its reverse side.

Figure 4.11 similarly shows coasters D, E and F. The pattern on coaster D is similar to coaster C, but has the castellations aligned. It is self-dual. The pattern on coaster E is jūjizashi (ten-cross stitch). The dual pattern has small squares and the stepped lines of *chiisana yamagata* (small mountain form). Coasters E and F feature dual designs.

The third set of coasters is shown in Figure 4.12. On coaster G there are aligned crosses, interspersed with aligned squares, a self-dual pattern. The pattern on coasters H and I is called *dan tsunagi* (linked steps). The steps run diagonally from the lower left to the upper right, and the only visual difference is a slight shift, whether the first step begins with the tread or the riser. Turning these coasters over, on a side edge or top or bottom, reveals linked steps that run 'the other way', from upper left to lower right. This pattern is also called dan tsunagi but in a reflected

Figure 4.11 From left to right, coaster D, coaster E and coaster F. Stitching and photograph by the author.

orientation. The front of coaster H and the reverse of coaster I are shown in Figure 4.13.

The duals of a number of other traditional stitching patterns are discussed by Seaton and Hayes (2023), and we recommended the following precise definition of the dual (p. 162):

> "... the dual pattern [is] what would be visible ... looking over the top of the work into a mirror placed behind the work. Then the stitches behind the left hand side of the work are on the left in the image in the mirror (and so too for right, top and bottom)."

Figure 4.12 From left to right, coaster G, coaster H and coaster I. Stitching and photograph by the author.

Figure 4.13 The front of coaster H and the reverse of coaster I; dan tsunagi in two orientations.

To obtain the stitching instructions for the dual, it is then only necessary to perform the exchange $0 \leftrightarrow 1$ in the binary strings that define the original pattern. We use the notation of a line above the binary string for this. For example, the vertical stitching lines for the dual side in *Duality Dyptych* in Figure 4.14 were obtained from those of *Stitch in Line* as

$$\overline{101101\ldots101} = 010010\ldots010 \ .$$

Referring to the strings in Table 4.1, this definition matches exactly what we know about the duality relationship of the pairs (B and C), and (E and F). We observed that they were the duals of each other, and we can see that the strings that generate them are related exactly by the $0 \leftrightarrow 1$ operation. Coasters H and I are also dual to each other with this careful definition of duality. This pair of coasters, which bear the same pattern albeit shifted slightly, also helps us to understand what is going on with A, D and G; although we've identified these patterns as self-dual, performing $0 \leftrightarrow 1$ on their strings seems to yield different, not the same, strings.

The strings indicate the presence (or absence) of the first stitch in a horizontal or vertical line of stitching, on the front of the work. If instead we chose to indicate the state of the *second* stitch, it would be in the opposite condition, absent or present. That is to say, the $0 \leftrightarrow 1$ operation can be thought of as shifting all vertical lines of stitching up (or down) by one stitch, and all horizontal lines of stitching left (or right) by one stitch. In general, this can change where and how the characteristic polygons and stepped lines form. Some patterns, the self-dual ones, are unchanged in general appearance by this, apart from an overall shift.

Figure 4.14 Our sampler *Stitch in Line* side-by-side with its dual form the piece *Duality Dyptych*. Stitching and photograph by the author.

4.4 MORE TO EXPLORE

- Stitch more pieces of hitomezashi (traditional or of your own design) and compare the 'front' with the 'flip side'. For pieces larger than the *Duality Coasters*, the effects around the edges on the back will be less noticeable. You could make your own dyptych of a pattern side-by-side with its dual.

- Explore dual polygons and dual polyhedra. For example, what is the dual of a rhombus? Of a kite? Of a cube? Of a tetrahedron?

- For the square (and the hexagonal) grid, the centre of a face is fairly intuitively where the perpendicular bisectors of the defining edges—or the diagonals—meet. Where is the centre of a triangle? There are actually a number of possible definitions, and although they coincide for equilateral triangles, this is not the case for isosceles or scalene triangles. Investigate the incentre, orthocentre, centroid and circumcentre.

- Draw the dual graph corresponding to the map coloured in Figure 2.5.

CHAPTER 5

Random Patterns

RANDOM is a word we use conversationally to describe things that we did not expect to encounter, or that seem to be irregular or to follow no discernible pattern. Hitomezashi gives us a way to experience what mathematicians mean by randomness, while at the same time creating aleatoric art.

5.1 SAMPLER: AT THE TOSS OF A COIN

Required materials:

- Coin or die or game spinner;
- Fabric, thread, needle and scissors; or grid paper, pencil and ruler.

Planning: For this one-of-a-kind non-traditional hitomezashi sampler, the state of each line of running stitch will be assigned at random. Decide first how many vertical and horizontal lines of stitching you will make. The example shown in Figure 5.1 uses 28 vertical and horizontal lines of hitomezashi stitching, each comprising 29 stitches and each determined by tossing a 10p coin, finished off with a double running stitch border.

Since there are two possible states for each line of stitching—beginning with a stitch on the top of the fabric (which we label as 1), or beginning with a stitch on the reverse (labelled as 0)—choose which of these states will correspond to your tossed coin coming down as a 'head'. Then the other state will correspond to tossing a 'tail'. Instead of a coin, you could use a game spinner, or a die with more (or fewer) sides than six; some possibilities are shown in Figure 5.2. If you use a die, rolling an odd number could specify the 1 state, and rolling an even number

DOI: 10.1201/9781003392354-5

Figure 5.1 A hitomezashi greetings card made for a relative during the separation of COVID-19. Created and photographed in 2020 by Jill Borcherds (UK) and used with her kind permission.

Figure 5.2 A selection of possible objects by which to generate random lines of stitching: coins, dice and spinner. Photograph by the author.

indicate 0. You could even use an online random number generator, or flip a virtual coin (see, for example, random.org [Haahr, 2023]).
Instructions: Whatever method you choose to make the random assignment, constructing this sampler proceeds by generating a 0 or 1, working the line of stitching, and repeating this until all your vertical and horizontal lines have been formed.

5.2 INCORPORATING CHANCE IN ART

Aleatoric art takes its name from the Latin word *alea* meaning dice. More generally, it refers to art that incorporates chance elements.

Not all indeterminacy arises in the same way. The twentieth century American composer John Cage (1912–1992) let the everyday sounds of the environment and the coughs and shuffles of the audience be the music in his famous (or infamous) 4′3″. When he 'prepared' a piano, he changed the sounds created by striking the keys, but in a way that would be the same each time a particular key was struck, until the piano was prepared afresh. By using the *I Ching*, a Chinese divination text of ancient origin, he achieved his goal of pure automatism, removal of the composer's free choices from the composition. Each of these is different in character from the control that a composer relinquishes when a performer is invited to improvise.

Mozart, and other composers of the eighteenth century, created musical games that involved chance. In his *Musikalisches Würfelspiel* (Musical Dice Game) he—or possibly his publisher—provided 'Instruction in the composition of as many waltzes as one desires with two dice, without understanding anything about music or composition' (this being the full title of the work, which is catalogued as K516f or K294d). Notation for 176 single bars of music were provided. With there being eleven possible outcomes of rolling two six-sided dice simultaneously, a sixteen-bar waltz could be created.

Kevin Jones (1991) explains how Mozart's game has been implemented with computers, including by John Cage: to render the dice rolls, to synthesise the instrument, and to incorporate further elements of randomness such as listeners fiddling with volume knobs or moving around. Jones himself is a composer of *stochastic music* in which random elements are generated according to strict mathematical laws.

In the visual arts, the mid-twentieth century saw the surrender of conscious control by abstract expressionists. Janet Sobel (1893–1968), whose work influenced Jackson Pollock, dripped and squirted paint from an eyedropper and sucked it across a horizontal canvas with a vacuum cleaner [Grovier, 2022]. Max Ernst experimented with decalcomania (pressing glass onto wet paint which created chance effects as it was lifted away) and oscillation painting (swinging a tin of paint with a hole in the base over a canvas) as in his work *Young Man Intrigued by the Flight of a Non–Euclidean Fly* (1942, 1947).

In her beautiful 2022 book *Record, Map and Capture in Textile Art*, Jordan Cunliffe proposes embroidery projects which incorporate elements beyond the stitcher's direct control: colour choice determined by the state of the sky, or stitch numbers corresponding to minutes of sleep. One project *Rolling Dice* requires a die to be rolled twice to determine each element of the piece. The first roll determines which of six small motifs to embroider, and the second how many times to stitch it. This is repeated until a possibly predetermined sufficiency of motifs has been worked.

Knitters, too, have used chance to create striped or lacy or textured fabric. Figure 5.3 shows part of the front panel of a child's sleeveless pullover in which the roll of a die determined the crossing of cables.

Figure 5.3 The position and direction (right over left or left over right) of cables in this detail from a child's pullover were determined randomly by rolling a die. Knitted and photographed by Morwenna Griffiths (Australia) c. 2014 and used with her kind permission.

This idea, courtesy of scientist Mary Griffin, appeared in the pages of *New Scientist* in 1987. It was the memory of the story behind this photo that prompted me to suggest creating aleatoric hitomezashi in the *Math Art Challenge.*

There is a sense of coming full circle in deliberately creating 'random' hitomezashi, rather than regular patterns. Functional lines of running stitch used to hold layers of cloth together for warmth or durability in the austerity of Edo period Japan would interact unintentionally in steps and loops, eventually giving rise to decorative forms [Hayes, 2019]. The full gallery description of the garment shown in Figure 1.2 compares boro to contemporary random collage.

5.3 PROBABILITY AND RANDOMNESS

Is your sampler *At the Toss of a Coin* really one-of-a-kind? Depending on the number of vertical and horizontal lines of stitches you made— let's call these numbers m and n respectively— there are 2^{m+n} different ways that your sampler could have turned out. With just five vertical and horizontal lines, that is over a thousand; with ten rows and columns, over a million! Jones (1991) estimates it would take 200 million years to play all of the waltzes that could be generated with Mozart's musical dice game.

To put it another way, if you (or someone else) constructed a second sampler of the same size, $m = n = 10$, by the same process, the probability of it turning out the same is about 0.000 001. It's actually greater than that, though still small, if we don't double-count samplers that are identical under rotation. We can remove 90° rotational symmetry from consideration by insisting that the sampler is rectangular, i.e., $m \neq n$. This does not eliminate symmetry under 180° rotation.

What would be the effect on our sampler if a single line of stitching was in a different state? If just one toss of the coin had been different? In Figure 5.4, one horizontal line has been changed compared to the greetings card in Figure 5.1. Can you spot which line it is? Because of the way that the vertical lines interact with the horizontal lines to form steps and loops that extend beyond single lines, you can see that some small loops have vanished, and there are longer runs of steps in the changed piece. (The line to change was chosen from among the 28 lines by using three dice: two twelve-sided and one four-sided. It's the 23rd from the bottom.)

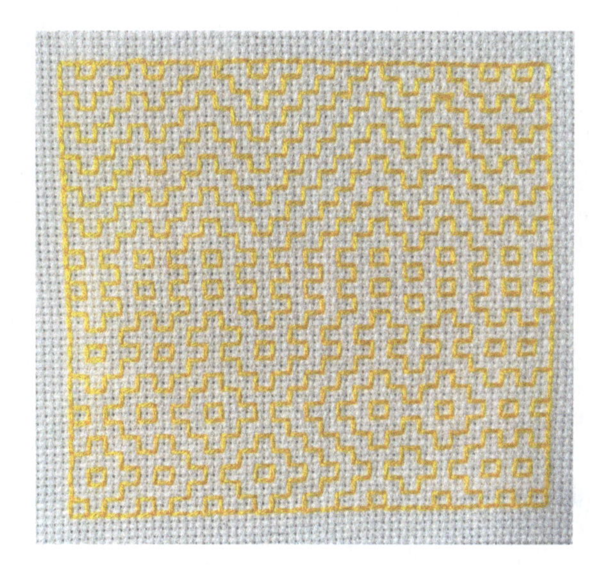

Figure 5.4 In recreating the piece shown in Figure 5.1, the state of a single line of stitching has been changed. Stitching and photograph by the author.

But why do we need to roll dice to generate a random pattern? Couldn't we just make it up? It turns out that our human attempts to be 'random' are not truly random. We distrust coincidence. Perhaps we are impatient; a balanced coin over a very large number of tosses will come up heads half the time, but we expect to see it happen in the short runs, too [Stewart, 2015]. Compare the two images in Figure 5.5. In creating the image on the left, the maker used her feeling for what would be random, and she has tended to avoid long runs of either 0 or 1, swapping to the other state after a run of two or three. However, the image on the right, which she generated by flipping a coin, contains some strings as long as five in a row of the same state.

Deviation from pure randomness can also be found in the seemingly haphazard works of painters such as Jackson Pollock; underlying the chance drips and pourings of his action painting there is balance, control and choice. The art critic Clement Greenberg said of Pollock's *Number 1, 1949* "Beneath the apparent monotony of its surface composition it reveals a sumptuous variety of design and incident" and analysis by conservators backs this up [Museum of Modern Art, 2013].

We should distinguish between the way we use the word 'random' conversationally, and a precise mathematical definition of randomness.

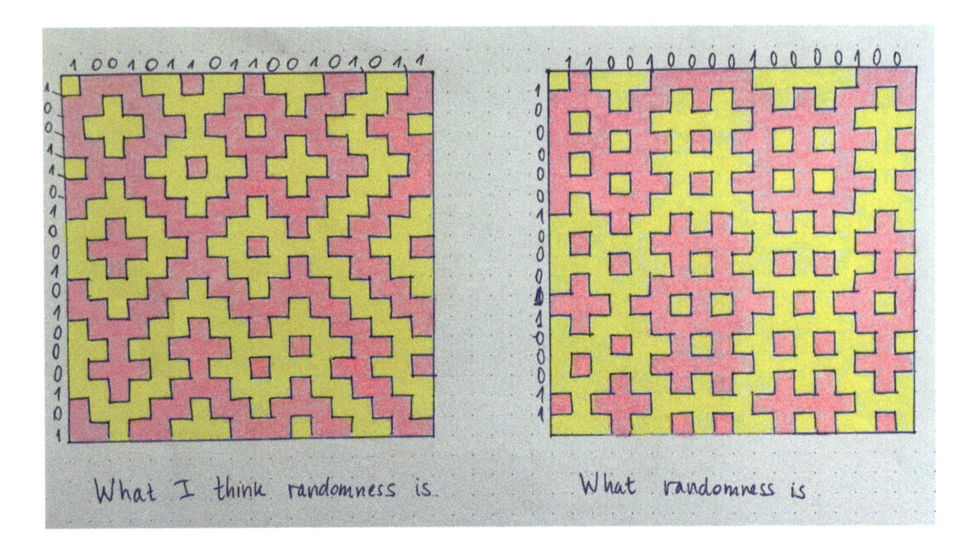

Figure 5.5 Random hitomezashi patterns drawn by 'feel' (on the left) and by using a coin flip (on the right). Created and photographed in 2020 by Constanza Rojas-Molina (France) and used with her kind permission.

We use it informally to mean irregular or messy or spontaneous or unexplained. We might even use chaotic as its synonym.

The mathematical definitions of random and chaos are tighter and distinct. Chaos is determined mathematically by what has already happened, and its unpredictability arises from sensitivity to tiny fluctuations in the starting conditions. A random outcome, on the other hand, does not depend on what came before it. A coin isn't influenced by the fact that the last time it was flipped, or the last five times, it came down heads. The result of a random experiment cannot be predicted, but a measure can be specified for how probable each of its possible outcomes is. This specification is called a *probability distribution*. It can be given as a formula, in a table, described in words, or graphically. We'll use tables and words.

The kind of probability distribution you used to create each line of stitching in *At the Toss of a Coin*, assuming that you used a balanced coin, is called a *Bernoulli distribution* with probability $\frac{1}{2}$. There are two possible things that could happen, either a head or a tail lands uppermost, and these each happen with equal probability. We mapped (or assigned) one outcome to 0 and the other to 1, so the states of the lines of stitching also follow a Bernoulli distribution with probability $\frac{1}{2}$. This is an example of a *one-to-one mapping*.

If we roll a single fair six-sided die, then each of the possible results has the same probability of appearing. There are six results, so they each have probability $\frac{1}{6}$. This probability distribution is an example of the *discrete uniform distribution.* More generally, this is the probability distribution for a list (discrete) of n possible outcomes, each having equal (uniform) probability $\frac{1}{n}$ of occurring. It is believed that the digits of π follow this distribution. (You might notice that the Bernoulli distribution with probability $\frac{1}{2}$ is the same as the uniform distribution with two outcomes.)

In Mozart's Musical Dice game the underlying probability distribution is non-uniform; the eleven possibilities for each bar of music do not all have the same probability of being selected. In Table 5.1, you can see the 11 possible totals for the number of dots uppermost when two six-sided dice are rolled simultaneously. Some of these can only happen in one way. For example, the total 2 only occurs when both rolled dice show a 1, which happens with probability $\frac{1}{6\times6} = \frac{1}{36}$. But others can happen in multiple ways. For example, the total 6 can arise from rolling 1 and 5 or 5 and 1, 2 and 4 or 4 and 2, or 3 and 3. These five results are mapped to the outcome that the total is 6. That is, it has probability $\frac{5}{36}$. The mapping of dice rolls to totals is an example of a *many-to-one mapping.*

If you created your sampler by rolling a fair six-sided die, and you mapped the six possible results to lines of stitching as suggested (odd numbers to 1 and even numbers to 0), this again gave equal probability for a line to be in state 0 or state 1. This situation is displayed in Table 5.2. But what happens if we map differently? We could map numbers with more than two factors (the composite numbers) to 0 and numbers with one or two factors (1 and the primes) to 1. Looking at Table 5.3, we see a non-uniform distribution, another Bernoulli distribution but this time with probability $\frac{1}{3}$ for one outcome, and thus $\frac{2}{3}$ for the other. The stitching states for the small piece *Random Sample-r* in Figure 5.6 were chosen in this way.

If you're getting a little tired of rolling dice, and all your coins have disappeared under the furniture (or you've joined the cashless economy), you might decide to turn to a computerised random number generator. Random number generators that use a mathematical algorithm to generate their output are in fact only pseudo-random; with enough information about the algorithm, the outputs can be predicted. This doesn't matter too much for designing a piece of embroidery. But for

TABLE 5.1 Outcomes and probabilities when rolling two six-sided dice

Total	Dice faces	Probability
2	1,1	$\frac{1}{36}$
3	1,2 ; 2,1	$\frac{1}{18}$
4	1,3; 3,1; 2,2	$\frac{1}{12}$
5	1,4; 4,1; 2,3; 3,2	$\frac{1}{9}$
6	1,5; 5,1; 2,4; 4,2; 3,3	$\frac{5}{36}$
7	1,6; 6,1; 2,5; 5,2; 3,4; 4,3	$\frac{1}{6}$
8	2,6; 6,2; 3,5; 5,3; 4,4	$\frac{5}{36}$
9	3,6; 6,3; 4,5; 5,4	$\frac{1}{9}$
10	4,6; 6,4; 5,5	$\frac{1}{12}$
11	5,6; 6,5	$\frac{1}{18}$
12	6,6	$\frac{1}{36}$

TABLE 5.2 'Odd and even' mapping of die result to stitching state

State	Die face	Probability
0	2; 4; 6	$\frac{1}{2}$
1	1; 3; 5	$\frac{1}{2}$

TABLE 5.3 'Composite number' mapping of die result to stitching state

State	Die face	Probability
0	4; 6	$\frac{1}{3}$
1	1; 2; 3; 5	$\frac{2}{3}$

Figure 5.6 This small piece called *Random Sample-r* has been generated by using a Bernoulli distribution with probability $\frac{1}{3}$. Stitching and photograph by the author.

cryptographic purposes something truly random, like atmospheric noise is better. This is how random.org generates its random numbers [Haahr, 2023] and it's how the piece shown in Figures 5.7 and 5.8 got its name.

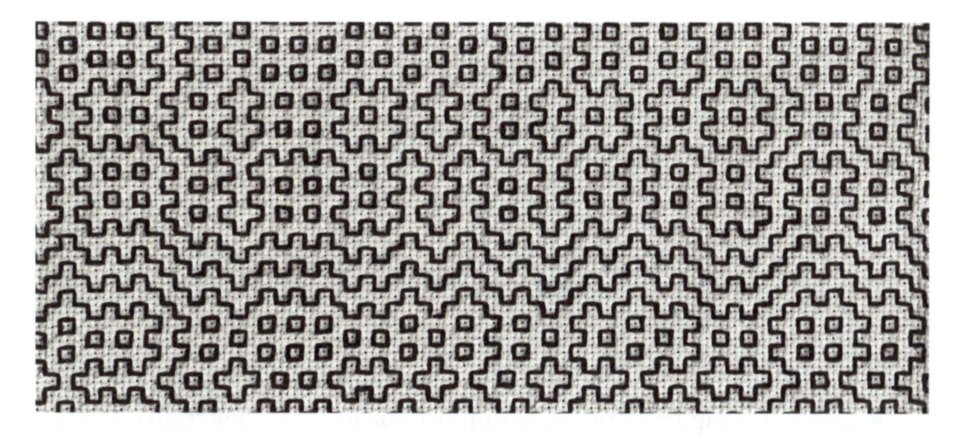

Figure 5.7 A detail from the aleotoric piece *Ever so Airy a Thread* (2023). Stitching and photograph by the author.

Figure 5.8 *Ever so Airy a Thread* (2023) consists of 110 lines of stitching in each direction, determined randomly. Stitching and photograph by the author.

5.4 MORE TO EXPLORE

- Choose a single line of stitching—perhaps randomly—and re-make your sampler *At the Toss of a Coin* changing the state of this line only.
- The line of stitching changed in creating Figure 5.4 was chosen by rolling three dice, two with twelve sides and one with four sides. Draw up a table of the probabilities corresponding to each possible total (from 3 to 28).
- Stitch a sampler using a non-uniform probability distribution, such as that in Table 5.3 and Figure 5.6.

- Watch the *Numberphile* video [Haran, 2021], in which mathematics communicator Ayliean McDonald draws hitomezashi patterns on square and isometric paper using randomness. She incorporates the digits of π and compares the effect of Bernoulli distributions with different probabilities.

CHAPTER 6

Symmetry

SYMMETRY in hitomezashi is perhaps its most visually obvious mathematical feature. In this chapter we consider which symmetries are present in traditional designs, and which rosette, frieze and wallpaper patterns are possible when stitching on the square grid.

6.1 SAMPLER: FRIEZE-R CONTAINERS

Required materials:

- Fabric, thread, needle and scissors; or grid paper, pencil, eraser and ruler.
- Seven jars or bottles or vases with a vertical (not tapered) surface.
- String or adhesive.

Planning: Plan which frieze pattern you will use for each of your containers. Measure the workable height of your various containers. Also measure the circumference. In using the stitching instructions in Table 6.1 the vertical string should be repeated as many times as it takes to go around the chosen container. However, because a frieze is a pattern that repeats only in one direction, the horizontal string is not repeated. The height of each frieze is indicated in Table 6.1.
Instructions: Cut paper or fabric to size for each container. Using the stitching instructions given as binary strings in Table 6.1 stitch or draw the patterns. The number of stitches to be worked in the vertical strings is indicated by the height given in the third column of Table 6.1. Where this height is greater (by one) than the number of horizontal lines specified, the vertical lines extend above and below the horizontal lines by one

DOI: 10.1201/9781003392354-6

TABLE 6.1 Stitching instructions for seven 'frieze-r' containers

Container	Vertical	Height	Horizontal
1	1010...	7 stitches	10100101
2	1010...	8 stitches	101010101
3	1010...	7 stitches	110101
4	101101...	5 stitches	101101
5	011011...	5 stitches	011001
6	101101...	7 stitches	01101101
7	1010...	8 stitches	101ϵ010

stitch. Where this height is less (by one) than the number of horizontal lines specified, the vertical lines fit inside the horizontal rows. Finally, there is a new symbol, the lunate epsilon ϵ, used for container 7. This symbol indicates an empty line, containing no stitches.

When specifying a hitomezashi pattern in this book, the symbol ϵ indicates a line empty of stitches.

To finish this project, either tie or attach the strips to the containers, as shown in Figure 6.1.

6.2 WHAT IS A SYMMETRY GROUP?

A mathematical *group* is a collection of objects, together with a method of combining them in pairs—a *binary operation*—in such a way that a set of four rules called the *group axioms* is obeyed. (We won't use details of these axioms.) For hitomezashi, we are interested in *symmetry groups*. The objects in a symmetry group are geometric transformations which do not change angles or lengths. In the two-dimensional plane, these *isometric transformations* are rotations, translations, reflections and glide reflections (and not dilations or shearing). A single glide reflection is comprised of a translation (the glide) and a reflection. The binary operation for these groups is to perform the first geometric transformation followed by the second; the result is still a geometric transformation, an object in the group.

Mathematicians can be a bit casual and talk about a geometric shape or pattern as being 'from such-and-such symmetry group', when really the geometric thing they are referring to is one that is visually

Figure 6.1 Five stitched, and two drawn, the wrappings around these jars illustrate the seven frieze groups. Stitching and photograph by the author.

unchanged—that is, maps onto itself—when the various transformations comprising that group act on it. We should say that the pattern or shape is symmetric under the action of the group, but we tend to just say that it 'has' that group's symmetries. We may even casually use the same label for the groups and the patterns.

There can be many geometric patterns that are symmetric under the same group. In our sampler *Frieze-r containers*, we stitched or drew one such pattern for each of the seven *frieze groups*. The *Almost infinite persimmon flower* of Figure 3.2 is symmetric under the action of one of the *point groups*. And Figure 6.2 shows a pattern of double and triple persimmon stitch that is symmetric under the action of one of the *wallpaper groups*—imagine extending the pattern by translation horizontally and vertically to cover an infinite wall, like wallpaper. We will examine each of these types of symmetry groups in more detail in the following sections.

The notation that we will use for the wallpaper and frieze groups and patterns is that of the International Union of Crystallographers (IUCr) also called Hermann-Mauguin notation, after its early proponents. For the rosette groups, we will use the notation of geometry for cyclic and dihedral groups. Other established notations exist, in particular the

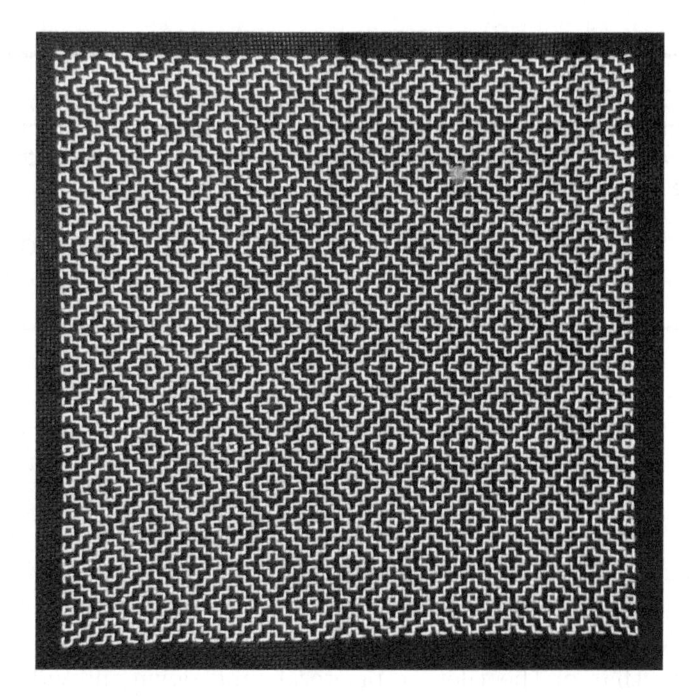

Figure 6.2 This traditional kakinohanazashi stitching is symmetric under the action of one of the wallpaper groups. Stitching and photograph by the author.

orbifold notation of Conway and Thurston (based on topological considerations), and Schoenflies notation for point groups in three dimensions.

6.3 SYMMETRY STUDIES AND SYMMETRY SAMPLERS

In their comprehensive book *Symmetries of Culture*, Washburn and Crowe (1988) propose using the mathematical classification of plane symmetries as a taxonomy for frieze and wallpaper patterns observed by archaeologists and anthropologists on cultural artefacts, such as cloth, pottery, basketry and carvings. Rather than the subjective impression given by subject matter and details of the motifs employed, they advocate for an objective and overall perspective, providing an introduction to one- and two-colour planar symmetries illustrated with numerous images of artefacts from around the world and across time periods. In particular, they suggest a fruitful framework for study would be to focus on what symmetries a culture selects or prefers. Crowe later wrote:

> "...it seemed to me that a comparison of patterns, perhaps by their presence or absence, or by the frequency of their occurrence, might give information about relations between cultures which were neighbors either in space or in time."
>
> Donald Crowe
> [Crowe, 2001, p. 2]

In a 2004 companion volume other voices joined this conversation. Carrie Bezine showed by demonstration that all 12 non-hexagonal one-colour wallpaper patterns can be woven as twill fabric using a four-shaft loom; she provided weaving drafts (instructions) for them. In another chapter it is pointed out that, because of the way that shuttle weaving progresses by laying down a complete line of weft across the warp threads before moving 'up' to the next, the mental challenge of designing woven patterns with particular symmetries is different from that of the tiler or printer laying down individual motifs [Franquemont and Franquemont, 2004]. The weaver works simultaneously on multiple motifs, but each pass of the shuttle adds only a small part to each. Roe and associates constructed designs using motifs based on those of a particular culture and then asked traditional artists to comment on these patterns. This made clear some unwritten rules, and insight into how, from all that is possible, certain patterns only are used or seem culturally 'correct'.

A somewhat different approach is taken by the numerous mathematical fibre artists who have produced symmetry samplers in techniques such as knitting, embroidery, bobbin lace, bead-crochet, temari and *shibori* dyeing. Their focus is on determining the largest possible range of symmetries which can be exhibited whilst conforming to the legitimate restrictions imposed by the chosen medium.

Samplers have been used historically as a learning tool for beginners in a craft, to provide a record of patterns intended to be shared, and sometimes to convey other information (numbers, the alphabet, platitudes) thus achieving two learning goals at once. Mathematical fibre art samplers showcase mathematical concepts together with the employed technique. Susan Goldstine gave *A Survey of Symmetry Samplers* in 2017, explaining how mathematical fibre artists have particularly sought to make *complete symmetry samplers* which show each possible symmetry consistent with the constraints of the chosen technique exactly once. Brezine's woven twill pieces, taken collectively, could be considered a complete symmetry sampler for four-shaft loom weaving.

Not only of mathematical or design interest, planar symmetries—in and of themselves, and as projections of three-dimensional symmetries—are important in crystallography. Pre-dating the mathematical fibre art literature are two papers of Hargittai and Lengyel, who used Hungarian needlework displaying the frieze and wallpaper patterns in the context of chemistry education. However, as the examples are drawn from several forms of embroidery, some constrained and others not, we would not categorise these compilations as complete symmetry samplers.

Carolyn Yackel has demonstrated all 14 three-dimensional spherical symmetry groups in temari (see Figure 1.6) [Yackel, 2011]. Bobbin lace is a medium in which all 17 wallpaper patterns can be constructed [Irvine and Ruskey, 2017]. These techniques place few constraints on the placement of threads.

Counted thread cross-stitch is constrained to lie on a square grid supplied by the fabric it is worked on. In 2007 Mary Shepherd created a piece called *Wallpapers in Cross Stitch*, which showed the 12 of the 17 wallpaper patterns consistent with the grid. More recently, Craig Kaplan added the 'missing' five wallpaper patterns, by stitching three-stroke asterisks on hexagonal weave cloth. However, as discussed by Baker and Goldstine (2012, 2015) in the context of bead-crochet, an underlying hexagonal grid excludes wallpaper patterns with four-fold rotation. Kaplan demonstrated the remaining 13; the *Crystallographic Bracelet Series* [Goldstine and Baker, 2011] shows 11.

Knitting stitches are rectangular, not square. Nine of the wallpaper groups, those compatible with a rectangular grid, can be demonstrated in the technique of double knitting; Susan Goldstine knitted them into a scarf and shared her pattern. Yackel (2021) has shown that exactly seven of the wallpaper patterns can be executed in Japanese shibori dyeing, by physically folding the cloth to be dyed in a variety of ingenious ways, to create the fundamental domains referred in the topological approach to symmetry of orbifold notation.

Wing Mui (2016) chose a *simple perfect square dissection* to display in blackwork all seven frieze groups (in one square), as well as the six rosette groups and 12 wallpaper groups consistent with the square grid upon which she stitched. A perfect square dissection is a dissection of a large square into smaller squares with integer-length sides, none of which are of the same size; it is called simple if the smaller squares taken together do not form any rectangles. In fact, the dissection Mui used was comprised of 21 squares, and she used the two smallest to display the empty set and a single point.

Cross-stitch is less constrained than hitomezashi in that the motifs are not forced to interact with each other. Due to the way that we construct hitomezashi with complete and correct lines of running stitch, tendrils of influence extend from the motifs (be they loops, steps or nested loops), locking the stitches above and below and on either side into particular states, akin to the challenge for the weaver. Another point of difference is that the stitches of hitomezashi outline regions on the plane, while cross stitches fill them.

Some fibre art realisations of symmetry are less than perfect. Beads are not usually rotationally symmetric and knitting stitches are shaped like a ‘V’. Cross stitches are square, but examined at the closest level (see Figure 4.8) there is an asymmetry because one of the two stitches making up each cross (say the SE to NW one) lies on top of the other (SW to NE). Generally, these minor asymmetries are politely ignored. Rendered on even-weave fabric, hitomezashi has the potential to have almost ideal symmetry because the stitches do not cross and they are all of equal length. However, we do see that when worked as counted thread technique on hessian, the inherent unevenness of the weave of the base fabric gives rise to crooked rows and deviation from a truly square grid. This is apparent, for example, in Figure 4.10.

Although worked in coloured thread and beads, all the designs mentioned thus far are mathematically one-colour patterns. Two-colour patterns, also called colour-exchanging patterns, are even more numerous: 17 frieze patterns and 46 wallpaper patterns. Yackel and Goldstine (2022) have demonstrated both theoretically and in physical form that 14 of the friezes and 22 of the wallpapers are attainable in mosaic knitting. Lorelei Koss has studied friendship bracelets, made in the knotted fibre art form macramé, both quantifying preference for particular symmetries and showing that all seven one-colour friezes, but not all two-colour friezes, can be knotted.

6.4 ROSETTES: SYMMETRY ABOUT A POINT

The point groups acting on a pattern on a plane are comprised of transformations that we can perform after we stick an imaginary pin into our pattern to create a fixed point. The pattern is anchored, so there can be no translations or glide reflections. If when we rotate the plane through

$\frac{360°}{n}$, where n is an integer, our pattern looks unchanged, lying exactly on itself after this rotation, we say it has an n-fold rotational symmetry about a point. Such points are also called rotation centres. The pattern can be finite. Alternatively it can extend theoretically indefinitely in all directions away from the fixed point.

If there are no reflection symmetries, only rotations through multiples of $\frac{360°}{n}$, the relevant symmetry group is the *cyclic group* C_n, with n elements. Patterns drawn or stitched on a square grid can have two- or four-fold rotational symmetry, or may have only the trivial full-circle rotation with $n = 1$.

The other possible point symmetry is reflection symmetry across one or more lines in the plane which go through the fixed point, the *reflection axis* or *axes*. This operation can only be pictured in our mind's eye, and not physically performed on a pattern—imagine a flat mirror placed along a reflection axis and perpendicular to the plane that the pattern lies in. What you can 'see' in the mirror corresponds to the 'other half' of the pattern. When there are two distinct reflection axes, reflecting in one followed by reflecting in the other results in a rotation. 'Dihedral' is an adjective which means 'relating to two planes'. Groups with reflections (and hence rotations as well) are called the *dihedral groups*.

The notation used for these groups in geometry is D_n and they have $2n$ elements. Polygons have this symmetry; the regular n-gon is unchanged if the group D_n acts on it.

An alternate picturesque name for the planar point groups is the *rosette groups*, because many flowers have one of these symmetries (approximately). The cherry blossoms *sakura* in Figure 6.3 have five petals and symmetry D_5. In the depiction of sakura by the small dish in Figure 6.4, the reflection symmetries are absent or broken. This dish only has rotational symmetry, so that its symmetry group is C_5. The persimmon flower as suggested by its sepals in Figure 3.3 has reflection axes and four-fold rotational symmetry (D_4).

Shown in Figure 6.5 is my first piece of hitomezashi to be exhibited—virtually, since it was in 2020—a complete symmetry sampler in which the six possible symmetry groups consistent with hitomezashi on the square grid are represented by a variety of traditional and modern hitomezashi motifs. Snippets from overall hitomezashi patterns have been selected. In most cases, the stitching does not extend to the edges of the area allocated for each rosette pattern, but the alternation of the running stitches is correct within each motif.

Figure 6.3 Cherry blossoms with five petals and approximate D_5 symmetry. Photograph by the author.

Figure 6.4 This stylised cherry blossom dish has approximate C_5 symmetry. Photograph by the author.

Figure 6.5 *Mrs Perkins' Persimmon Quilt* (2020). Stitching and photograph by the author.

I called this piece *Mrs Perkins' Persimmon Quilt* as a deliberate homage to the square dissection used for the remarkable sampler of Mui. Mrs Perkins' quilts take their name from a puzzle, at least 100 years old, involving Mrs Potipher Perkins who set herself the challenge of unpicking some of the seams in a square patchwork piece she was given for Christmas, to create smaller square quilts. That is, they are dissections of a square into smaller squares, some of which may be the same size as each other, and all of which have integer length sides. There can be more than one Mrs Perkins' quilt of each original size. Chosen for this sampler is a dissection of a 3×3 square into six squares (5 unit squares and one of size 2×2), there being six rosette patterns to depict.

The largest motif (a many-layered persimmon flower acknowledged in the name of the piece) is symmetric under the action of D_4, with the eight elements:

- Rotation through 90°, 180°, 270° and 360° about the central fixed point;

- Reflection through four axes that run through the fixed point; two aligned with the grid of the fabric and two at 45 degrees to it.

Below this large motif, the motif comprised of four offset jūji (ten-crosses) has two reflection axes aligned with the fabric grid, and has 2-fold rotational symmetry. It is symmetric under D_2.

Also below the large motif is a motif with a single vertical axis of symmetry and its only rotation being 360°. In an echo of the work of Roe, this motif struck Carol Hayes when she saw it as being not particularly Japanese. Indeed, I devised it based on what was possible to realise the D_1 representative and to fill as much of the space allocated for it in the quilt as possible, whilst providing visual contrast with the various stepped lines in other of the motifs.

In the lower right-hand corner, the traditional hitomezashi igeta motif is surrounded by a frame of stitching drawn from a shōnai sashiko design resembling counting rods (*sankuzushi*). The framing breaks the reflection symmetry at the corners and the relevant group is C_4.

Above this is a section of the pattern jōkaku judiciously cropped to have only 2-fold rotation. Look closely at the final stitches in each row to see that there are no reflections, so that it is symmetric under C_2. As Susan Briscoe notes, patterns can be known by more than one name, and this name is taken from her 2022 book; elsewhere I have called it *hirayama michi* (passes into the mountains) based on another source.

In the top right corner is a motif with no symmetry apart from rotation through a full 360°, so that it represents symmetry (so to speak) under C_1. This 'motif' is a square cropped out of an infinite persimmon flower, like that in Figure 3.2.

In achieving this symmetry sampler of the rosette groups consistent with hitomezashi stitching rules and the underlying square lattice, precise truncation of overall patterns into motifs was required for C_1 and C_2. In the case of C_4, the frame is made of running stitch, but there are vertices away from the edge which have degree one only, not degree two as they should if they were truly stitched according to the rules we have set ourselves for hitomezashi.

6.5 FRIEZE PATTERNS

Frieze patterns run along a line, theoretically forever in both directions, comprised of translated copies of a base unit. They are a common decorative feature in architecture and textiles and around the edges of vessels.

There are exactly seven frieze groups. The IUCr denotes them using four symbols:

- a first symbol gives information about the repeating unit (always p[rimitive] for friezes).
- the second symbol denotes the highest order of rotation present (2 if there is 180° rotation, otherwise 1).
- the third symbol indicates if there is a reflection axis perpendicular to the direction in which the frieze runs; m indicates a reflection axis (mirror) exists and 1 is used in this place if not.
- the fourth symbol indicates either a reflection (m) or a glide reflection (g) parallel to the direction in which the frieze runs, and 1 indicates neither is present.

We will illustrate these symmetry groups by looking at a piece created during the winter of 2020, which I called *You must be friez-ing in that.* Access to shops in Melbourne (arguably the world's most locked-down city) was not possible and cut-to-length specialised cloth could not be purchased for click-and-collect. I found a vintage child's pyjama top (mine in the 1970s) in my fabric box, featuring polka dots which could provide a grid for my stitches.

Single zig-zag lines of mountain form *yamagata* divide the stitching on the back, shown in Figure 6.6, into three sections, each of them a frieze. The pattern at the top consists of a row of kakinohanazashi offset from a row of crosses. There is vertical reflection symmetry as well as translation. This pattern is symmetric under the frieze group p1m1.

Along the bottom runs a frieze of persimmon flowers which has horizontal and vertical reflection symmetry, as well as translation and 180° rotation. The vertical reflection axes are between the motifs, and running through the centre of each of them. The rotation centres lie on these axes. The frieze group associated with this pattern is p2mm.

The pattern in the middle (two offset rows of crosses) has translation, vertical reflection, 180° rotation and glide reflection. The associated frieze group is p2mg. The rotation centres are a bit harder to spot, lying between the arms of adjacent crosses.

In Figure 6.7, which shows the front of the pyjama top, the pattern at the top left as it faces us has translation symmetry only, hence being symmetric under the action of the group p111 or just p1.

Figure 6.6 Three friezes from a complete symmetry sampler of the frieze groups *You must be friez-ing in that* (2020). Stitching and photograph by the author.

Figure 6.7 Four friezes from a complete symmetry sampler of the frieze groups *You must be friez-ing in that* (2020). Stitching and photograph by the author.

Below it is stitched the pattern dan tsunagi with translation and 2-fold rotation, the associated frieze group being p211 or just p2.

The pattern on the top right is yamagata (with the mountain peaks on their side) and it has a reflection axis parallel to the direction in which the frieze runs. This pattern is labelled p11m. Below it is a pattern in which a horizontal line of stitches has been left empty. This empty line provides the axis for a glide reflection, and the symmetry is that of p11g.

The three patterns on the back of the pyjama top all use the same pattern of vertical lines of stitching (010 repeated), and the four on the front are based on vertical lines in alternating states. As well as the truncation technique utilised previously to realise the rosettes and used here to create a p1 pattern, the new ploy of leaving a line of stitches empty has been introduced. Both of these defects can be seen in pictures of traditional pieces of hitomezashi, whether due to inattention or shortage of thread (missing lines of stitches) or running out of space (truncation).

6.6 WALLPAPER PATTERNS

A wallpaper pattern consists of motifs that are repeated by translation in two directions. The overall pattern can have symmetries that the individual motifs do not have, as we will see. The IUCr labelling for the wallpaper patterns is similar to that for friezes:

- a first symbol gives information about the repeating unit (p[rimitive] or c[entred]).
- the second symbol denotes the highest order of rotation present.
- the third symbol indicates if there is a reflection or glide reflection axis perpendicular to the main translation direction; 1 is used in this place if not.
- the fourth symbol indicates either a reflection (m) or a glide reflection (g) parallel to the main translation direction (or at an angle other than a right angle with it); 1 indicates neither is present.

Seventeen wallpaper patterns exist, but five of them involve rotations of order 3 or 6, not compatible with the square lattice. How many of the remaining 12 can we demonstrate in hitomezashi?

When I started my first wallpaper piece (see Figure 6.8) in 2020—and it is intentionally on the scale of a piece of wallpaper, at 79 cm

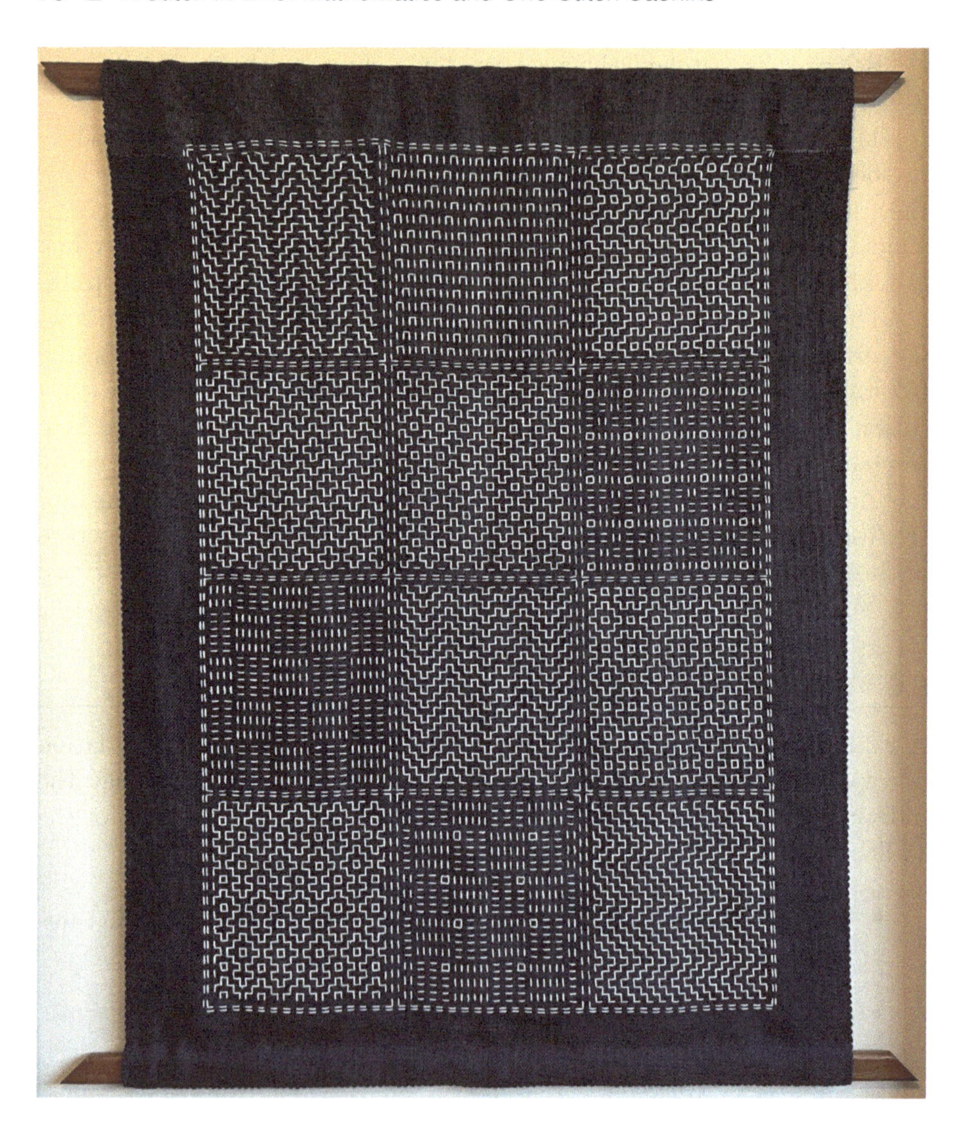

Figure 6.8 *Hitomezashi Wallpaper* (2021). Stitching and photograph by the author.

by 104 cm—as a beginner, and being strongly influenced by traditional hitomezashi designs, I could think of no way to execute several of the wallpaper patterns (specifically p111, p1g1, p4gm and p411) without using incomplete rows of stitches. That is, not empty rows, but short sections or individual stitches that were placed predictably but which required the thread to be cut or to be carried across the back of the work between them. This was a source of frustration to me, and later in this section some of the wallpaper patterns are illustrated with designs that I have recognised more recently, and which do not bend the rules we've imposed on our hitomezashi.

The placement of patterns on *Hitomezashi Wallpaper* was chosen to give visual balance between those which contained loops such as persimmons and crosses, those with stepped lines, and those having vertices of degree one which appear less dense. Another way to arrange a complete symmetry sampler is by working through the groups systematically based on common features. We will follow more closely the latter organisation as we discuss the patterns, beginning with some examples that help with understanding the distinction between patterns with primitive and centred units.

Consider the two patterns shown in Figures 6.9 and 6.10. They each have 2-fold rotation symmetry, and have horizontal and vertical reflection axes. Their labelling will have the form ?2mm. The pattern in Figure 6.9 has its rotation centres at the intersections of pairs of reflection axes. The pattern in Figure 6.10 has rotation centres at such intersections, but also in between, in this case on the stepped lines. The pattern in Figure 6.9 is p2mm (or pmm for short) and that in Figure 6.10 is c2mm (or cmm for short). Textile designers recognise 'c' patterns when one of the translations is what they term a 'half-drop'.

Continuing with patterns which have highest order of rotation 2, it's straightforward to identify horizontal reflection axes in the pattern in the top right-hand corner of Figure 6.8. This pattern, which is not traditional insofar as I have been able to determine, also has vertical glide reflection axes. The rotation centres lie on these vertical glide reflection axes. The appropriate label is p2mg (shortened to pmg).

A second glide reflection can be introduced by incorporating empty stitching lines, as seen in the top left-hand corner of Figure 6.8. Although only vertical stitching lines are empty, there are both horizontal and vertical glide reflection axes, and the pattern is p2gg (pgg). The rotation centres lie off the glide reflection axes.

The final pattern with 2-fold rotation is p211 (or p2), which has only rotations, not reflections. In hitomezashi, this can be realised by using mountain forms that rise and fall by a different number of stitches. The pattern in the bottom right-hand corner in Figure 6.8, in which the mountain forms are again on their side, has this symmetry.

There are four wallpaper patterns with only the trivial 360° rotation: p1m1, c1m1, p1g1 and p111.

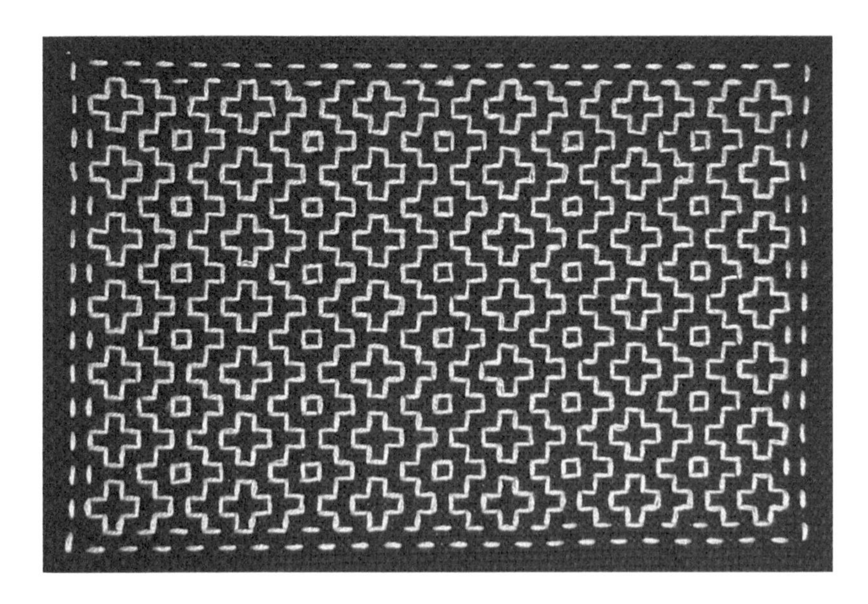

Figure 6.9 Hitomezashi symmetric under the action of p2mm. Stitching and photograph by the author.

Figure 6.10 Hitomezashi symmetric under the action of c2mm. Stitching and photograph by the author.

Figure 6.11 The pattern on the right is invariant only under translation. The pattern on the left, wherein strategic vertical lines of stitching are omitted, also has glide reflections. Stitching and photograph by the author.

To demonstrate a pattern with reflection axes in one direction only, p1m1 (pm), I again used the idea of steps that rise and fall by different amounts in such a way as to remove any rotations. The pattern in the middle of the third row from the top in Figure 6.8 is of this type; only a single repeat (horizontally) has been stitched.

Consider the two images in Figure 6.11. The pattern on the right, another variant on mountain form, is of type p111 (p1); translation is the only transformation under which there is symmetry. But by removing some vertical lines of stitching from this design, as shown on the left, a pattern has been created that has vertical glide reflection axes: p1g1 (pg). (Neither of these was used in the original *Hitomezashi Wallpaper.*)

The pattern in the bottom left-hand corner of Figure 6.8 has a half-drop translation and vertical reflection axes only; this pattern is of type c1m1 (cm). (Together, the half-drop and vertical reflections give rise to glide reflections.) This is not a traditional pattern, being rather a fusing of persimmon and cross; personally, I think it resembles a flower on a stalk with two small leaves.

The persimmon design in Figure 6.2—which has both triple persimmon flowers *sanju kakinohanazashi* and double persimmon flowers *nijū kakinohanazashi* offset from each other—is an example of wallpaper pattern type p4mm (p4m). That is, there are rotation centres for 4-fold rotations, and reflection axes in perpendicular directions. These fundamental symmetries lead to further glide reflections, reflection axes at 45° to the underlying square grid, and rotation centres for distinct 180° rotations. (There are, unintentionally, two examples of this wallpaper pattern in Figure 6.8.)

We have demonstrated 10 of the 12 wallpaper patterns compatible with the square grid in hitomezashi. For two of them (only) empty lines of stitching have been used to introduce glide reflections without reflections; while most vertices in the stitching have degree two, there are admittedly vertices of degree one within these two patterns.

The remaining two wallpaper patterns have four-fold rotations, being p4gm (p4g) and p411 (p4). The Mirror Property of hitomezashi loops mentioned in Chapter 3 gives reason to believe that these patterns cannot be executed even by bending our self-imposed rules to the extent we have so far.

The pattern used to exemplify p4gm in *Hitomezashi Wallpaper* resembles wickerwork *ajiro* found on bamboo screens or ceiling panels. Although it appears to be made with interwoven short rows of stitches, it can be stitched with an overall movement from edge to edge, working blocks of running stitches on the square grid perpendicular to the direction of movement as shown by the red stitches in Figure 6.12. To

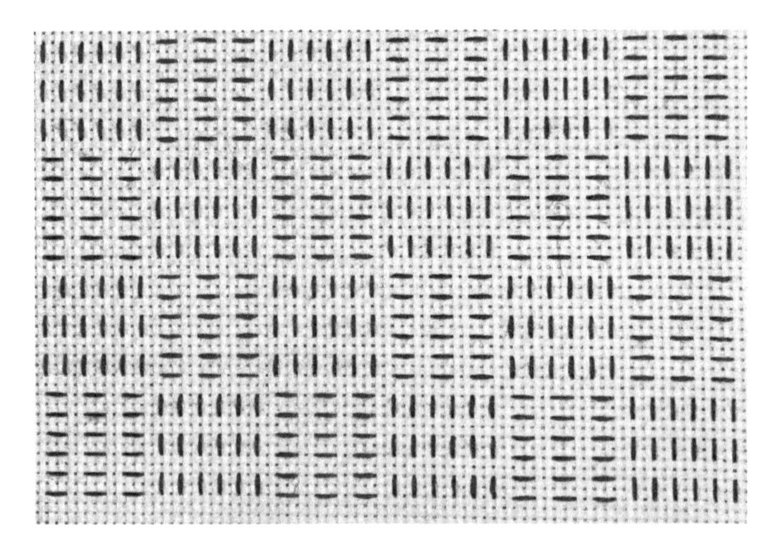

Figure 6.12 Stitching symmetric under the action of p4gm. Stitching and photograph by the author.

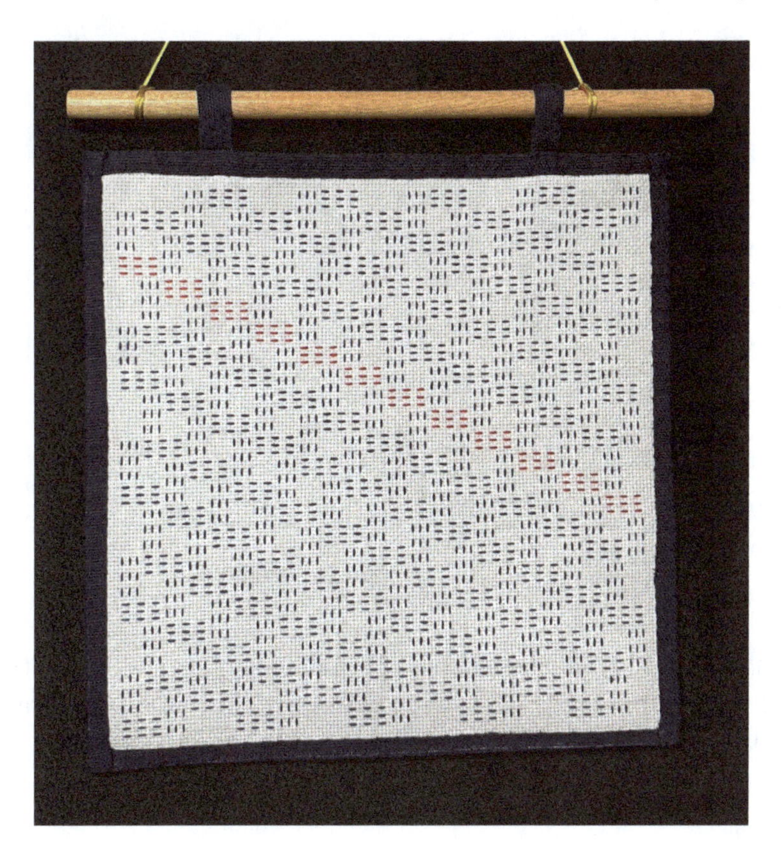

Figure 6.13 Stitching symmetric under the action of p4. Stitching and photograph by the author.

step from one small block to the next, one diagonal stitch is used on the reverse. Threads do not need to be cut or carried.

Ajiro features in the names of a variety of sashiko stitch patterns which resemble weaving, some of which incorporate empty space. After many pieces of grid paper ended up on the floor, I devised the design shown in Figure 6.13. Again the red stitches show how it can be worked moving from edge to edge, never carrying or cutting the thread. The reverse (of the incomplete piece) is shown in Figure 6.14. This design has 4-fold rotation as its only symmetry, and hence realises the wallpaper pattern for p4.

Neither of these latter two patterns can be expressed in terms of strings of 0 and 1 (and ϵ) in our usual manner. Instructions are given for the other wallpaper patterns discussed above in Table 6.2.

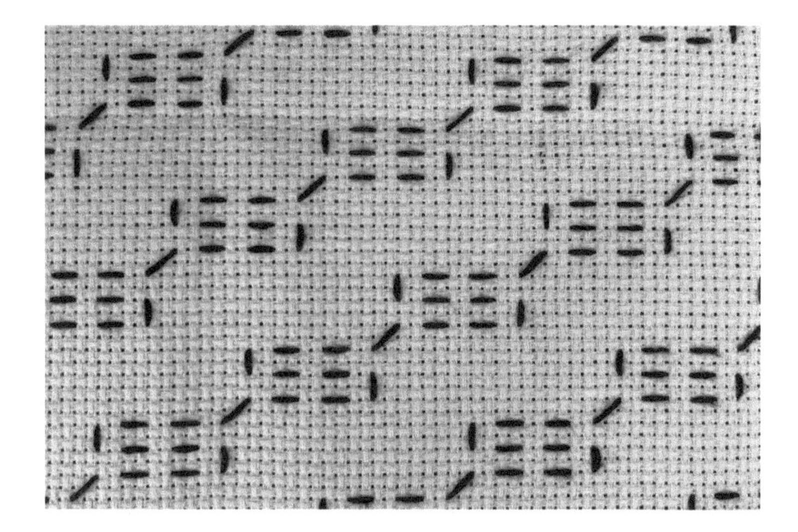

Figure 6.14 Detail from the reverse of stitching symmetric under the action of p4, photographed while in progress. Stitching and photograph by the author.

TABLE 6.2 Hitomezashi stitching instructions for representative wallpaper patterns

Pattern	Vertical	Horizontal
p1m1	01	10100100011000100101
c1m1	101	100011011100
p111	11001110	10
p2mm	0110	01101001
c2mm	0110	01101
p2mg	01101001	0100
p211	01	10101101
p4mm	0110	0110
p2gg	ϵ101ϵ010	10
p1g1	ϵ100ϵ110	10

6.7 SYMMETRIES IN HITOMEZASHI PATTERNS

The completeness of symmetry samplers provides a nice introduction to the limitation of existence proofs in mathematics. Constructing an example is sufficient to prove that something exists. Non-existence is subtler. Just because you and I can't think of a way to demonstrate it right now, that does not guarantee that it could not be done with more creativity or insight! I experienced this coming back to *Hitomezashi Wallpaper* to write about it some three years after I stitched it. Not only did I find that one pattern was missing (and one was doubled-up), I realised that two more patterns could be stitched correctly in hitomezashi in ways I had not thought of before. A well-argued justification is required in order to convince ourselves, and others, that a particular thing cannot be.

If we confine ourselves by insisting that we use fully packed lines of stitching (that is, all internal vertices have degree exactly two) then we have seen exhibited:

- Five of the six rosettes compatible with a square grid; not C_4.
- Six of the seven frieze patterns compatible with a square grid; not p11g.
- Eight of the twelve wallpaper patterns compatible with a square grid; not p4gm, p411, p2gg and p1g1.

Permitting empty lines of stitches adds the remaining frieze and two more wallpaper patterns (p1g1 and p2gg). That is, it has enabled glide reflection in the absence of a reflection, but not four-fold rotation without two reflections.

6.8 MORE TO EXPLORE

- The seven *Frieze-r containers* numbered 1, 2..., 7 have designs similar, but not all identical, to those on the pyjama top. Associate to each of your containers the appropriate frieze group label.
- Look back at the samplers *Stitch in Line* and *The Great Sine Wave*. Under which frieze group is *The Great Sine Wave* symmetric? Taken as a whole *Stitch in Line* has p1m1 symmetry. Can you identify smaller sections of it with different symmetry, as for the back of the pyjama top?

- Look back at the nine *Duality Coasters.* Under which wallpaper group (or rosette group) is each of them symmetric?
- There are many other possible hitomezashi patterns which demonstrate the various point, frieze or wallpaper symmetries explored in this chapter. Design and execute some of your own, or make some pieces using the instructions given in Table 6.2.
- Some of the stitching designs shown in this chapter are not traditional, so far as we know. Can you suggest suitably evocative names for them?
- Something to think about: do a hitomezashi design and its dual always have the same symmetries?

III

Generating Art

CHAPTER 7

Codes, Coding and Algorithms

PARALLELS between needlework and computing have frequently been drawn in the context of knitting and weaving. In this chapter, we see that hitomezashi has the potential to contain secret messages and hidden writing, encoded in the binary strings that specify the lines of stitching. We also consider it as a medium for algorithmic art.

7.1 SAMPLER: WHAT'S IN A NAME?

In her Numberphile video about hitomezashi drawing, Ayliean McDonald converts a six word phrase and digits of π into a piece of generative art [Haran, 2021]. We'll start with something smaller than that.

Required materials:

- Fabric, thread, needle and scissors; or grid paper, pencil, eraser and ruler.
- Writing implement and paper for planning.
- Optional: metallic or silk thread, backing fabric such as felt, a brooch back (available from the jewellery-making section of haberdashery shops) or a safety pin.

Planning: On your planning paper, write one word horizontally, and another word vertically. You could use your name, or choose any two other words that have personal significance for you and which can be written in an alphabet with consonants and vowels. Beside each consonant write a 1, and beside each vowel write a 0.

DOI: 10.1201/9781003392354-7

Figure 7.1 A hitomezashi brooch *What's in a name?* is shown pinned onto a knitted scarf (c. 1980). Stitching, knitting and photograph by the author.

Instructions: Use the two binary strings as the stitching instructions for a small piece of hitomezashi. In Figure 7.1 you can see the piece stitched using my given and family names, finished off in the form of a brooch. I've trimmed the fabric I stitched on, folded the unstitched parts to the reverse side, and then sewn a rectangle of felt cut to size onto the back using whip stitch. The tiny piece has been converted into a wearable, personalised piece of art by attaching a brooch back. If you're planning to wear yours, consider using some special silk or metallic thread for the stitching. If you draw your piece, perhaps you could use it for a bookmark.

Reflection: What's in *your* name? I was delighted to find most of a persimmon flower in mine. It's personalised, but not exclusive to me. The same piece of stitching could have been produced by someone named, say, Wendolesa Cooper! We've created stitching instructions which give a pleasant result by mapping from a familiar alphabet to the binary alphabet of 0,1 but not in a way that is unique or that can be reversed.

7.2 SPEAKING IN CODE

There was once a time when the word code meant a set of rules—and it was only used as a noun, not as a verb—and there was a time when a computer was a person who performed computations (think of the women in the movie *Hidden Figures*). We're not living in those times, so let's spend some time exploring these words and some related terms.

There are two meanings, or shades of meaning, of 'code' that we might use when talking about needlework.

One meaning is that of a secret code, or a way to encrypt (make secret) a message, more accurately called a *cipher.* When authorities banned the sending of knitting patterns overseas during the second world war, in case they contained sensitive information in code, it was this meaning they had in mind.

Morse code, named for the American painter and inventor Samuel Morse (1791–1892), provides the rules to convert letters into a form that can be transmitted as long and short pulses of sound or light or electricity. Originally created to send messages over long distances by telegraph, it is not per se a secret code. When in a war movie a spy is seen sending a message in Morse code, he or she has probably first enciphered or encrypted it. *Cryptography,* changing the original message into a form unintelligible except to an authorised recipient, requires another set of rules, those for the cipher. Our spy would encrypt the message, and then encode it into Morse code and transmit it. The intended recipient would need knowledge of the decryption rules to decipher the message, once decoded from dots and dashes into letters, into meaningful form.

Charles Dickens did not specify how his fictional character Madame Defarge in *A Tale of Two Cities* (originally published in 1859) enciphered the names of those doomed to destruction within her knitted register. They are described as "[k]nitted, in her own stitches and her own symbols". When a concerned fellow revolutionary asks about deciphering this information, he is assured that it is "as plain to her as the sun".

But when someone says today that knitting is [like] coding they probably mean that knitting instructions strongly resemble the code in which computer programs are written, and knitters approximate a machine as they execute them (or vice versa, since knitting predates machines). By following these instructions with yarn as input, the knitter-as-processor produces fabric as output. Most knitting involves two basic stitches K (knit) and P (purl), just as instructions for computers are expressed using two digits, 0 and 1. Cate Kennedy's poem *Binaries* (originally published as *Relative Complexities* in 1996) is a delightful commentary on this parallel.

In Figure 7.1, my hitomezashi brooch is pinned to a scarf, knitted in a pattern called Mistake Rib. There is no literal mistake; the squishy textured fabric has instructions very similar to those for regular ribbing but worked on an odd rather than even number of stitches. The knit stitches look like a V, while the perpendicular bumps correspond to purls. The instructions for Mistake Rib are:

Cast on a multiple of 4 stitches, plus 3 more stitches.

Row 1: *K2 P2; repeat from * to the last three stitches, K2 P1.

Repeat row 1 until the desired length is reached.

Bind off all stitches.

This resembles a computer program or sub-routine, incorporating 'if' statements and 'while' loops [Howard, 2017].

Now that we understand encoding as being the use of rules (the code) to change information or instructions into another form, we can see that beginning in Chapter 2 we've encoded the information required to stitch hitomezashi patterns into binary strings, drawn from an alphabet of two symbols, 0 and 1. We've also carried out a form of *compression*, which is computer-speak for reducing the resources required for the storage or transmission of information. The terse form of knitting patterns with abbreviations and implied loops developed as a way to save space in pattern books and magazines. We've already noted how compact our hitomezashi instructions are, compared to a chart or the finished piece.

Another term from the world of computing that has found its way into the language of fibre art is the term 'hack'. Not meaning to chop one's knitting into pieces (the word for that is steek), it is used by machine knitters who program their knitting machines by way of a personal computer to produce designs and items never originally in their remit. Just some of the many fibre artists working in this way are Madeleine Shepherd and Sarah Spencer.

In a sense we've traced another full circle, then. The ingenious Jacquard attachment, which used punched cards to control warp threads or bobbins in the mechanical weaving looms and lace machines of the Industrial Revolution, inspired Babbage and Lovelace in their envisaging of a Universal Analytical Machine. From the mechanical calculating devices of the nineteenth century, by way of the electro-mechanical Bombe built for code-breaking at Bletchley Park, and the electronic computers that took up entire rooms and operated in parallel with human computers in the space program of the nineteen-sixties, to today's personal electronic devices with integrated circuits on tiny silicon chips, we arrive back at textiles!

Cryptography means secret writing; it conceals the meaning of a message, but not that a message exists. *Steganography* means hidden writing, disguising the fact that there is a message at all. A fibre-art example of steganography is a cross-stitch sampler created in a prisoner-of-war camp by Major Alexis Casdagli during World War Two. Hidden in plain sight, dots and dashes which appeared to his captors to be ornamental were stitched in the border of the piece. These dots and dashes in fact comprised patriotic messages in English, rendered in Morse code. There was encoding but no encryption of the messages.

The brooches or drawings we created in the sampler *What's in a name?* are not encryptions, since they can't be unambiguously decrypted. We used a many-to-one mapping from one alphabet to another. How, then, might we use hitomezashi for hiding a message that only the cognoscenti could read?

Jordan Cunliffe devotes a third of her book *Record, Map & Capture in Textile Art* to steganography. At first sight, coloured beads and stitches are a delight for the eye. Only the initiated know that her personal diary entries are on public display. Many of her items use ASCII (the American Standard Code for Information Interchange), employed in computer programs, which in its extended form assigns a number to each of 256 characters (including letters, numerals, punctuation marks and letters with diacritics). These 256 numbers can in turn be expressed as binary numbers. By converting text to ASCII, and ASCII to binary, and using the binary strings as hitomezashi instructions, we can hide a message in plain sight. We need two such strings, one for the vertical lines of stitching and the second for the horizontal.

7.3 SAMPLER: STEGANOSTITCHING

In our next sampler, you can hide your own message in plain sight. In Figure 7.2 I have used the given and family names of a friend; her names happen to both have five letters so this piece is square. In Figure 7.3 I have used the names of a bride and a groom, making them a card for their wedding. For both pieces I used only the five digits that come after the three-digit prefix common to all capital letters in ASCII. You could instead use all eight digits which enables you to distinguish lower case letters from capitals, and to incorporate letters with diacritics or numerals and punctuation.

Figure 7.2 The given and family names of a friend encoded in binary, by way of ASCII, provide the stitching instructions for this piece. Stitching and photograph by the author.

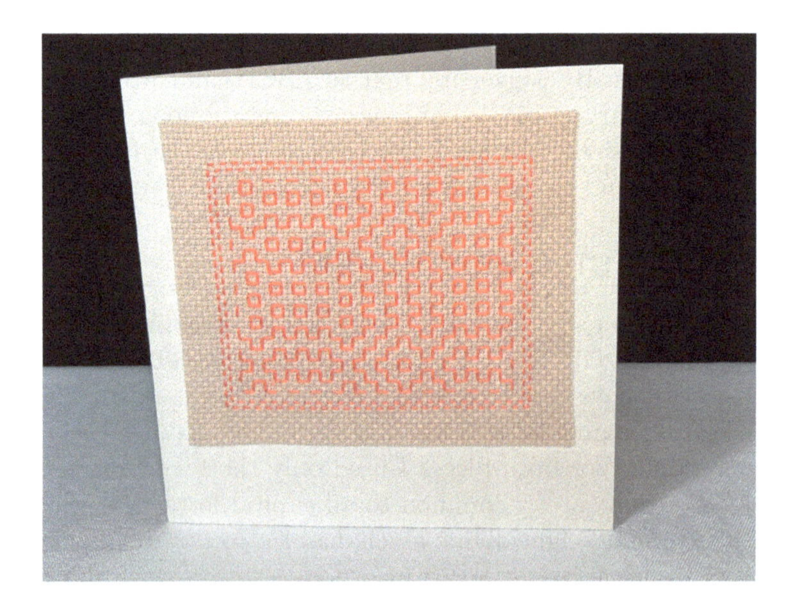

Figure 7.3 Wedding card, featuring the names of the bride and groom in steganostitching. Stitching and photograph by the author.

Required materials:

- Fabric, thread, needle, and scissors; or grid paper, pencil, eraser and ruler.
- An ASCII-to-binary table (most conveniently looked up online).
- Writing implement and paper for planning.

Planning: Choose two words or phrases to hide in plain sight. As you choose your words, remember that each character becomes eight binary digits or eight lines of stitching. Use the ASCII table to convert the letters in these phrases into binary strings.

Instructions: Now use the two binary strings as the stitching instructions for a piece of hitomezashi. It is easy to lose track of the irregular pattern of zeroes and ones, so I recommend transcribing the binary strings onto paper and ticking each one off as you complete stitching it.

7.4 ALGORITHMIC FIBRE ART

What can we say then about the brooch made in the sampler *What's in a name?* While it does not contain a decipherable message, it is an example of algorithmic fibre art.

An *algorithm* is a list of unambiguously stated steps to be followed. It leaves nothing open to interpretation. (A recipe, on the other hand, is often less precise, containing phrases such as 'use sufficient water'.) A closely related idea is that of generative art, art in which the control is handed from the artist to an autonomous system. While algorithmic and generative art can be produced digitally, they need not be. These styles fall under the wider umbrella of conceptual art, where the creative emphasis is on choosing or defining the concept.

The weaver and mathematics teacher Ada Dietz (1888–1981) used an algorithm based on the expansion of algebraic expressions such as $(x+y)^3$ to determine new drafts for weaving. Her designs were recognised as truly original by other weavers. She said of them:

"As patterns grew and possibilities opened up, I found that mathematics gave the beautiful space divisions, proportions, and individuality of pattern which the artist strives to achieve."

Ada K. Dietz
[Dietz, 1949, p. 2]

Having developed a mathematical model for bobbin lace, Veronika Irvine employed a backtracking algorithm to search systematically and exhaustively for potential new bobbin lace background or filling designs, not all of which she has presumably made (yet), there being over 10^6 of them [Irvine and Ruskey, 2014, Irvine, 2015].

Madeleine Shepherd and Julia Collins proposed the use of Lindenmayer systems or L-systems to generate organic-looking knitted and crocheted pieces in their participatory project *Botanica Mathematica*. Lindenmayer (1925–1989) was a Hungarian biologist, based in the Netherlands, who modelled cell growth. In an L-system, a string of symbols grows by applying production rules to its constituent symbols. Shepherd and Collins interpret the symbols as coloured crochet or knitted stitches, and each application of the production rules generates a new row of knitting or round of crochet. Both in *Botanica Mathematica* and in a machine-knitted Möbius scarf exhibited at the Bridges conference in 2018, cellular automata have been used to generate intriguing two-coloured patterns [Matsumoto et al., 2018]. Shepherd also draws on Conway's Game of Life (a two-dimensional cellular automaton) for fairisle designs. Joshua and Lana Holden (2021) provide an extensive survey of the use of cellular automata in the fibre arts.

Although non-Euclidean geometry is the mathematics on display in the *hyperbolic crochet* of Daina Taimiņa, its creation is algorithmic: crochet $n-1$ stitches, and increase by one stitch in stitch n. These instructions followed exactly give rise to a crocheted surface with constant negative curvature. Figure 7.4 shows both an L-system flower and some hyperbolic crochet.

Figure 7.4 A piece of hyperbolic crochet (on the left) and a knitted L-system flower. Knitting, crochet and photograph by the author.

In a similar vein, *sequence knitting*, introduced by Cecelia Campochiaro (2015) and studied by Sara Jensen (2023) is algorithmic. Both Cunliffe and Campochiaro find the use of an algorithm in needlework allows mindfulness, presence in the moment without the pressure of furiously thinking ahead about the next design decision.

Just as Campochiaro developed new knitted fabrics, Irvine new bobbin-lace fillings and Dietz original weaves, in subsequent chapters we will find new hitomezashi patterns by applying an algorithmic approach to specifying the binary strings for our lines of stitches.

7.5 MORE TO EXPLORE

- Knowing that the eight-digit ASCII codes have been truncated to five digits by dropping the three-digit prefix, can you decipher the names in Figures 7.2 and 7.3? (In Figure 7.2, the first name runs along the top, and the second name down the right-hand side.)

- Make more pieces of hitomezashi steganostitching, or algorithmic brooches or bookmarks or greeting cards for friends.

- Morse code is a variable-length code, with the most frequently used letters having the shortest encoding. Can you think of a way that Morse code could be used to create hitomezashi-like stitches? Not only do the long and short pulses need to be considered, it is also necessary to know where the encoding for one symbol stops and the next starts.

- Read the transcript of Kris Howard's talk at the Linux conference in 2017 for more details of how knitting is like coding [Howard, 2017].

- Explore the possibilities of Ada Dietz's algebraic weaving designs at the Wolfram Demonstations Project [Zeleny, 2013].

CHAPTER 8

Snowflakes

FIBONACCI snowflakes are the topic of this chapter, the most technical in the book. The reward for the technicality is the plethora of beautiful, fractal hitomezashi patterns, many of which appear to be new, specified by using recurrence relations to generate binary strings to serve as the stitching instructions.

8.1 SAMPLER: SNOWFLAKE BUNTING

The bunting shown in Figure 8.1 was created to meet a challenge set by Katie Steckles in *The Aperiodical* in 2020, to create *Fractal Bunting.* The instructions given below have been refined to give identically sized flags and better symmetry.

Required materials:

- Fabric, thread, needle, and scissors; or grid paper, pencil, eraser and ruler
- Ribbon or similar to display the bunting.

Planning: The stitched area of each flag consists of 16 vertical and horizontal lines of stitching, and if you would like to work a simple border (as in Figure 8.1), allow for two extra lines in each direction when calculating the size of fabric to use. Because my flags were worked on hessian, which does not have an even weave, they appear rectangular rather than square. They've been finished very simply—unhemmed and pinned to ribbon. Allow more fabric if you wish to hem them and sew them to a strip of fabric. Instructions are given for six distinct flags. For a longer string of bunting, make more than one of each.

DOI: 10.1201/9781003392354-8

Figure 8.1 A string of *Snowflake Bunting.* Stitching and photograph by the author.

Instructions: For each flag, the vertical and horizontal lines are worked following the same instructions, given in Table 8.1.
Reflection: Turn over the flags and look at the reverse. What do you notice about their symmetry and their duality?

8.2 RECURRENCE RELATIONS FOR BINARY STRINGS

The name of Fibonacci (who lived in Italy at the turn of the thirteenth century) is used to label a sequence of numbers which were known to Indian mathematicians centuries earlier. The nth Fibonacci number can be generated by the following second-order *recurrence relation*

$$F_{n+1} = F_n + F_{n-1}, \quad n \geqslant 1 \tag{8.1}$$

given two initial numbers $F_0 = 1$ and $F_1 = 2$. Then

$$F_2 = F_1 + F_0 = 2 + 1 = 3; \quad F_3 = 3 + 2 = 5; \quad F_4 = 5 + 3 = 8$$

and so on: 1, 2, 3, 5, 8, 13, 21, 34.... The Fibonacci numbers are well-known, including in popular culture, and can be used to count many things, including (at least approximately) things in the natural world.

TABLE 8.1 Stitching instructions for *Snowflake Bunting*

Flag label	Stitching instruction
21	0110011001100110
22	1100111001110011
$\overline{22}$	0011000110001100
41	0110110110110110
42	0100110110110010
$\overline{42}$	1011001001001101

We'll note now, for future reference, that the numbers F_1, F_4, F_7 and in general F_{3k+1} (for $k=0,1,2,\ldots$) are even.

By choosing different values for F_0 and F_1 we could generate an entirely different sequence of integers using the same recurrence relation. (Actually, sequences that start with $F_0=0,\ F_1=1$ or $F_0=1,\ F_1=0$ or Fibonacci's choice $F_0=1,\ F_1=1$ each settle down after the first few terms to give essentially the same sequence as we have above.)

A different second-order recurrence relation generates the integer sequence called the *Pell numbers*

$$P_{n+1}=2P_n+P_{n-1}, \quad n \geqslant 1 \tag{8.2}$$

given two initial numbers $P_0=0$ and $P_1=1$. The first seven Pell numbers are

$$0,\ 1,\ 2,\ 5,\ 12,\ 29,\ 70.$$

Again, these numbers were known in antiquity; the naming for John Pell (an English mathematician of the seventeenth century) is well established but known to be a misattribution. Fibonacci did at least pose a famous problem related to the sequence to which his name has been attached! He described the musings of a person having

---

"...one pair of rabbits together in a certain enclosed place, and one wishes to know how many are created from the pair in one year when it is the nature of them in a single month to bear another pair, and in the second month those born to bear also."

Fibonacci
From a modern translation [Sigler, 2003, p. 404]

Just as we can generate the next in a sequence of integers by addition of the two preceding terms, we can use an analogous recursive process to create a family of binary strings, which in this context are often called *words on a two-letter alphabet.* We'll use the symbol f_n for the nth *Fibonacci word*, formed by concatenation of the two preceding words in analogy to Equation (8.1):

$$f_{n+1}=f_n f_{n-1}. \tag{8.3}$$

We must choose two initial words to start with, f_0 and f_1. The first six Fibonacci words are shown in Table 8.2 for the choice $f_0=0$ and $f_1=01$.

TABLE 8.2 Fibonacci words with binary complement and reverse

n	f_n	$\overline{f_n}$	$\widetilde{f_n}$
0	0	1	0
1	01	10	10
2	010	101	010
3	01001	10110	10010
4	01001010	10110101	01010010
5	0100101001001	1011010110110	1001001010010

The length of the Fibonacci words is given by the Fibonacci numbers: $|f_n| = F_n$.

There are two things that we will want to do to binary words in the context of defining hitomezashi stitch patterns. If we use the symbol w for an arbitrary word or string, the notation $\overline{w}$ indicates the *binary complement*, that is, making the exchange of $0 \leftrightarrow 1$ for each letter in the word. We used this transformation when we discussed dual patterns in Chapter 4. We use the notation $\widetilde{w}$ to indicate the *reverse* of the word. They are both *involutions*: the original word is recovered if either of these transformations is done twice. The result of performing these transformations on the first six Fibonacci words is shown in Table 8.2.

A word w is a *palindrome* if it is the same written forwards and backwards: $\widetilde{w} = w$. We call w an *antipalindrome* if taking both the binary complement and writing the word backwards gives the original back: $\overline{\widetilde{w}} = w$. Among the Fibonacci words, f_1 is an antipalindrome, and f_2 is a palindrome.

Transformations that can be done proceeding letter-by-letter are called *substitutions* or *morphisms* because by this means the word morphs into another word. Forming the binary complement is a morphism. Reversal however is not, because we can't carry this process out by application of a substitution rule. To write down the reverse of a word, we need to be able to see the whole word at once.

Monnerot-Dumaine (2009) described six different morphisms that can be performed on the basic Fibonacci word in order to generate drawing rules on the plane. The one relevant for hitomezashi generates patterns that consist solely of left and right turns. It can be performed on the words of even length, f_{3k+1}. Monnerot-Dumaine first defined the *dense Fibonacci words*: take the letters in the Fibonacci word in pairs. The possible pairs are 00, 01, and 10 (because 11 never occurs in a Fibonacci word). Treat these as numbers written in base two, and convert

TABLE 8.3 Fibonacci words transformed to drawing instructions by way of dense Fibonacci words (DFW)

n	f_n	**DFW**	**Drawing rule**
1	01	1	R
4	01001010	1022	RLL
7	$f_6 f_5 f_6$	10221022110211021	$RLLRLLRRLRRLR$

them to decimal notation: $00_2 = 0$, $01_2 = 1$ and

$$10_2 = 1 \times 2^1 + 0 \times 2^0 = 2.$$

These are then interpreted as drawing rules for a path on a square lattice. Begin facing east (say). Then 0 means do nothing, 1 means turn right R and take one step, and 2 means turn left L and take one step. The first few dense Fibonacci words and the associated drawing rules (with 'do nothing' suppressed) are given in Table 8.3. The resultant drawings are shown in Figure 8.2.

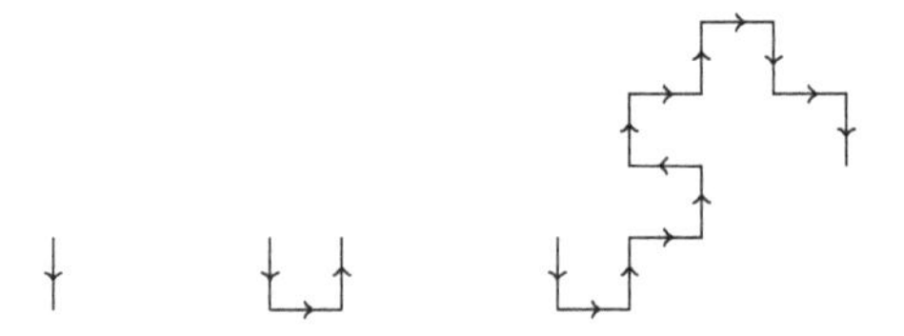

Figure 8.2 Implementation of the drawing rules based on Fibonacci words by way of the dense Fibonacci words. Created by the author.

8.3 FIBONACCI SNOWFLAKES

Draw a line segment (like the one on the left in Figure 8.2), and make three more copies of it. Rotate one copy through 90°, one through 180° and one through 270°. Joining them end-to-end gives a square. Now look back at the square grid (Figure 4.3); each edge is shared by two faces, so that every square face has four neighbours. Searching systematically for polyominoes that tile the plane in a similar way to squares, Blondin Massé and collaborators found many such *pseudo-square* tilings, including a beautiful family which they call the *Fibonacci tiles* or *Fibonacci snowflakes* [Blondin Massé et al., 2011]. The derivation relies on words similar, but not identical, to the words defined by Equation (8.3). The resultant polyominoes have as their boundary four copies of the paths

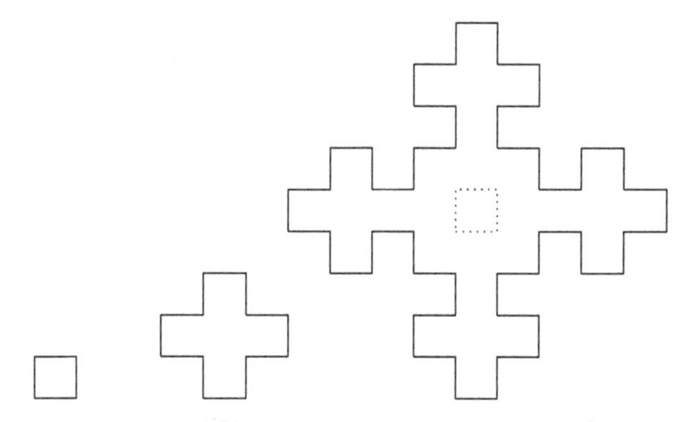

Figure 8.3 Drawings of the first three Fibonacci snowflakes. Created by the author.

obtained from the Fibonacci words by the drawing rule of Monnerot-Dumaine.

So, let's take the other two drawings from Figure 8.2, make three more copies of each and rotate one copy through 90°, one through 180° and one through 270°. Joining them end-to-end gives us two more Fibonacci snowflakes, as shown in Figure 8.3. We immediately recognise two of them as traditional hitomezashi motifs, kuchizashi and jūjizashi. The line segments forming the outline of the third one, and necessarily the small square indicated at its heart, can also be stitched this way, using only horizontal and vertical lines of running stitch on the grid. They feature on three of our *Snowflake Bunting* flags.

The lengths of the paths (in grid steps) are given by Fibonacci numbers which are odd. Thus the total perimeter length is, as we know it should be for a hitomezashi loop, congruent to 4 modulo 8. The height (and width) of the first few Fibonacci snowflakes are: $1, 3, 9$ and their areas are $1, 5, 29$. We have only a few examples here, but in general the widths can be expressed in terms of Pell numbers as $2P_k - 1$, and the areas are $P_k^2 + P_{k-1}^2 = P_{2k-1}$. We call k the *order* of the Fibonacci snowflake. These properties are collected in Table 8.4. You may like to check that the order four Fibonacci snowflake shown in Figure 8.4 has the specified properties.

The crosses are pseudo-squares, that is, they can tile the plane in such a way that each is surrounded by four copies of itself, as indicated in Figure 8.5. The higher order, more crenellated Fibonacci snowflakes fit together like intricate puzzle pieces to tile the plane. These close-packed

TABLE 8.4 Fibonacci snowflake properties

Drawing rule	Perimeter	Area	Width	Order
RLL	12	5	3	2
RLLRLLRRLRRLR	52	29	9	3
	$4F_{3k-4}$	P_{2k-1}	$2P_k - 1$	k

tilings are not potential hitomezashi patterns, because there are vertices of order four.

8.4 PELL PERSIMMON POLYOMINO PATTERNS

Examining the *Snowflake Bunting* flags suggested to me how snowflakes could be laid out in a pattern that can be stitched as hitomezashi. Worked in running stitch, motifs must have exactly one grid spacing between them. The flags labelled 21 and 41 show traditional designs, which are self dual. The largest motifs are aligned in rows and columns, with smaller motifs appearing between and within them. The stitching instructions are well known and easily converted to our notation in binary strings, as we have seen in earlier chapters. But if the entire family of Fibonacci snowflake outlines can be defined using a recurrence

Figure 8.4 Fibonacci snowflake of order four, detail from a larger piece *Cross and Crown* (2022). Stitching and photograph by the author.

Figure 8.5 Fibonacci snowflakes as pseudo-squares, indicated using the order 2 case. Created by the author.

relation for words, I wondered if was possible to do something similar to generate hitomezashi stitching instructions. The answer appears to be yes [Seaton and Hayes, 2023].

A key observation is that the width of the Fibonacci snowflakes, in stitches, is one more than the width counted in grid squares: $2P_k$. They have D_4 symmetry, so that a word of length P_k is sufficient to generate the order k motif, when used together with its reverse. The proposed word should agree with the cases already stitched, and generate a self-dual pattern. I experimented with a number of different concatenations of binary strings based on Equation (8.2) before defining the *Pell words* for $k \in \mathbb{N}$ by:

$$p_k = \overline{p_{k-1}}\ \overline{\widetilde{p_{k-2}}}\ p_{k-1} \tag{8.4}$$

with p_0 being the empty word and $p_1 = 1$. The length of these words is $|p_k| = P_k$. To have the correct reflection symmetry, the hitomezashi stitching pattern is specified by

$$w = p_k \widetilde{p_k}.$$

Using this word as the instruction for both the vertical and horizontal stitching produces a pattern which is self dual. There is one such pattern for each $k \geqslant 1$. But because the Pell numbers grow exponentially, to physically see the self-duality of patterns for higher values of k the piece one makes has to be quite large, at minimum $4P_k$. Details taken from *Cross and Crown* is shown in Figure 8.4 and 8.6. A single-order six Fibonacci snowflake is shown in Figure 8.7, which is 140 stitches wide. Here overstitching of the largest motif in blackwork makes it visible. The piece gets its name *Jubilee*—and its colour scheme—from the length of the $k = 6$ Pell number, 70; I was working on it at the time of Queen Elizabeth II's platinum jubilee.

Figure 8.6 Fibonacci snowflake of order five, detail from a larger piece *Cross and Crown* (2022). Stitching and photograph by the author.

Figure 8.7 Hitomezashi piece: *Jubilee* (2023). Stitching by the author; photograph by Dan Bach (USA) and used with his kind permission.

Following the established instructions, the overall piece is symmetric under the action of the group p4mm. Snowflakes of lower order appear around and inside the largest snowflakes because of the recursive definition (Equation 8.4) which generates the pattern. To distinguish these designs from a single snowflake as generated by a walk following the instructions of Blondin-Massé et al., or a tiling of the plane by juxtaposed snowflakes as in Figure 8.5, I call these hitomezashi designs the *Pell persimmon polyomino patterns.* This name, incorporating the words Pell and persimmon, captures the four-fold symmetry and the Pell numbers which characterise the polyomino motifs.

It's actually quite a bold assertion, that this construction of hitomezashi stitching patterns, based on widths, symmetry and observations of a few low order cases, will keep generating Fibonacci snowflakes for all higher values of k. Like any mathematician putting forward a general claim, I tested it against some higher order patterns not used in proposing it. Gradit and Van Dongen have subsequently tested it to extremely high order. In mathematics, a good-looking claim awaiting a proof is called a *conjecture.*

Persimmon-Snowflake Conjecture
The largest polyomino in the Pell persimmon polyomino pattern of order k is the Fibonacci snowflake of order k.

A number of other polyominoes which are pseudo-squares are illustrated by Blondin-Massé et al. (2010). Not all can be stitched as hitomezashi. Those that can include all the persimmons which we associated with the centred square numbers in Chapter 3. Ramírez and co-authors gave the specification for i-generalised Fibonacci words and snowflakes using walks and we will give hitomezashi instructions for them too.

But first we will consider a question that may have occurred to you—how, when natural snowflakes have six-fold symmetry, did the Fibonacci snowflakes with their four-fold symmetry get their name?

8.5 FROZEN FRACTALS

Some seventy years before Mandelbrot gave us the word fractal, and before computers, in 1904 a Swedish mathematician Helge von Koch introduced a curve obtained by iterating a geometric substitution

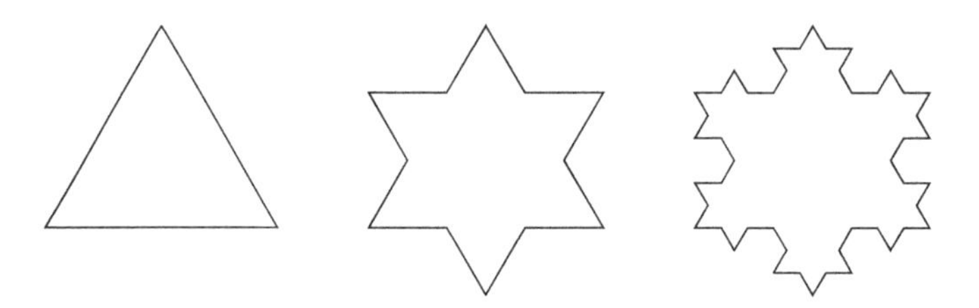

Figure 8.8 The first three Koch snowflakes. Created by the author.

process. Applying this process beginning from an equilateral triangle, as in Figure 8.8, a series of spiky objects with six-fold symmetry emerges.

The object obtained by repeating these steps indefinitely is called the (von) Koch snowflake, and has properties of complexity and self-similarity viewed at any scale. (The various stages of iteration are also often referred to casually as Koch snowflakes.) Complexity and self-similarity are the hallmark of fractals. Being generated by an iterative or recurrence process is often a feature of mathematical fractals, though not so obviously of naturally occurring fractal objects, such as coastlines, fern leaves or broccoli. These are all features of the Fibonacci tiles and hence their naming as snowflakes, analogously.

Indeed, observers have looked at my pieces such as *Jubilee* and have remarked—mistakenly—that I have stitched some Koch snowflakes. I've been asked on other occasions whether I have stitched a space-filling curve such as the Hilbert or Moore curve. (No, there are many finite loops.) The persimmons have been mistaken for Aztec diamonds. I've never been asked if I've stitched the second Sierpiński curve, also called the Sierpiński square curve; it's not Sierpiński's best-known fractal. The resemblance here is stronger (see Figure 8.9) but that curve cannot be stitched using hitomezashi running stitches.

Figure 8.9 Second iterate of the Sierpiński square curve. Created by the author.

The word 'fractal' carries with it for mathematicians the idea of a dimension, appropriately defined, which is fractional (non-integer) between the dimension of the curve (one) and of the space where the fractal lives (two). The fractal dimension of the Fibonacci snowflakes is

$$\frac{\log(2+\sqrt{5})}{\log(1+\sqrt{2})} \approx 1.638.$$

The *golden ratio* ϕ associated with the Fibonacci numbers, makes a disguised appearance: $\phi^3 = 2+\sqrt{5}$, while $1+\sqrt{2}$ is the silver ratio, associated with the Pell numbers [Blondin Massé et al., 2012].

Mention fractal art and most people think of digital art, computers being employed to carry out thousands of iterations. However, traditional woven and embroidered textiles from many parts of the world feature the scaled self-similarity and complexity that characterise fractals. The Sierpińksi triangle and the Sierpiński carpet have been incorporated directly into contemporary fibre art either through intentional holes [Wildstrom, 2008] or textures [Bernasconi et al., 2008].

8.6 SNOWBALLING

Ramírez et al. defined i-generalised Fibonacci numbers and words and Pell numbers, by changing the initial values or strings used in the recurrence relations Equations (8.1), (8.2) and (8.3). The associated drawing rule, once again, gives only the outline of one generalised snowflake. For even values of i these are achievable following our hitomezashi stitching rules. We will call i the *type* of the number, word or pattern, and write it as a square-bracketed superscript (to distinguish from the order, still written as a subscript, and from a power).

The i-generalised Fibonacci numbers (see Table 8.5) are

$$F_n^{[i]} = F_{n-1}^{[i]} + F_{n-2}^{[i]}$$

with $F_0^{[i]} = 1$ and $F_1^{[i]} = i$ for natural numbers $n \geqslant 2$ and $i \geqslant 1$.

TABLE 8.5 Generalised Fibonacci numbers, for $n = 0, 1, 2, \ldots$ and i even.

$F_n^{[2]}$	1, 2, 3, 5, 8, 13, 21, ...
$F_n^{[4]}$	1, 4, 5, 9, 14, 23, 37, 60, ...
$F_n^{[6]}$	1, 6, 7, 13, 20, 33, 53, 86, ...

TABLE 8.6 Generalised Pell numbers, for $n = 0, 1, 2, \ldots$

$P_n^{[0]}$	0, 1, 2, 5, 12, 29, 70, ...
$P_n^{[1]}$	1, 2, 5, 12, 29, 70, 169, ...
$P_n^{[2]}$	2, 3, 8, 19, 46, 111, ...
$P_n^{[3]}$	3, 4, 11, 26, 63, 137, ...
$P_n^{[4]}$	4, 5, 14, 33, 80, 199, ...

For $i = 0, 1, 2, \ldots$, define i-generalised Pell numbers:

$$P_{n+1}^{[i]} = 2P_n^{[i]} + P_{n-1}^{[i]}, \quad n \geqslant 1 \tag{8.5}$$

with $P_0^{[i]} = i$ and $P_1^{[i]} = i+1$. The case $i = 0$ corresponds to Equation (8.2). Lists of these generalised Pell numbers are given in Table 8.6.

We want to modify the initial words by concatenating them with 0 and 1, depending on the type i. Because we only use even i, we can employ this neat trick:

$$\frac{1+(-1)^{\frac{i}{2}}}{2} = \begin{cases} 1, & i = 0 \bmod 4 \\ 0, & i = 2 \bmod 4 \end{cases}$$

and similarly

$$\frac{1-(-1)^{\frac{i}{2}}}{2} = \begin{cases} 0, & i = 0 \bmod 4 \\ 1, & i = 2 \bmod 4 \end{cases}.$$

Then building on the initial words from the $i = 0$ case, we define generalised initial words using a recurrence relation in the type (not the order). Beginning from $p_0^{[0]} = \epsilon$

$$p_0^{[i]} = \frac{\left(1-(-1)^{\frac{i}{2}}\right)}{2} p_0^{[i-2]}$$

and from $p_1^{[0]} = 1$,

$$p_1^{[i]} = \frac{\left(1+(-1)^{\frac{i}{2}}\right)}{2} p_1^{[i-2]}.$$

The first few of these are given in Table 8.7.

The words of higher order are then generated using the analogue of equation (8.4):

TABLE 8.7 Initial words to generate generalised Pell words

Type	$p_0^{[i]}$	$p_1^{[i]}$
$i = 0$	ϵ	1
$i = 2$	1	01
$i = 4$	01	101
$i = 6$	101	0101
$i = 8$	0101	10101

$$p_k^{[i]} = \overline{p_{k-1}^{[i]}}\ \overline{\widetilde{p_{k-2}^{[i]}}}\ p_{k-1}^{[i]}. \tag{8.6}$$

The lengths of the words are given by the generalised Pell numbers: $|p_k^{[i]}| = P_k^{[\frac{i}{2}]}$. Again, working analogously, the stitching instructions for hitomezashi are specified in each direction by

$$w = p_k^{[i]}\widetilde{p_k^{[i]}}$$

for $i = 2, 4, 6, 8, \ldots$ and $k = 1, 2, 3, \ldots$. A few of these are given in Table 8.8. The patterns for $i = 2$ are not visually distinct from the 'original' Pell persimmon polyomino patterns—this was a source of no little confusion as I attempted to sort out the technical details!

The width of the type i order k snowflake is $2P_k^{[\frac{i}{2}]}$ in lines of stitches, and thus $2P_k^{[\frac{i}{2}]} - 1$ in grid squares, which is odd (as we know it must be). Ramírez et al. showed that its area is

$$\text{Area} = \left(P_{n-1}^{[\frac{i}{2}]}\right)^2 + \left(P_n^{[\frac{i}{2}]}\right)^2.$$

We observe (and we could prove in a straightforward way) that the numbers in Table 8.6 alternate in parity in each row. Thus in any consecutive

TABLE 8.8 Stitching instructions for some generalised Pell persimmon polyomino patterns

Type	**Order**	**Word** w
$i = 2$	$k = 3$	011100110001
$i = 4$	$k = 3$	1011001001001001101
$i = 6$	$k = 2$	10100100101

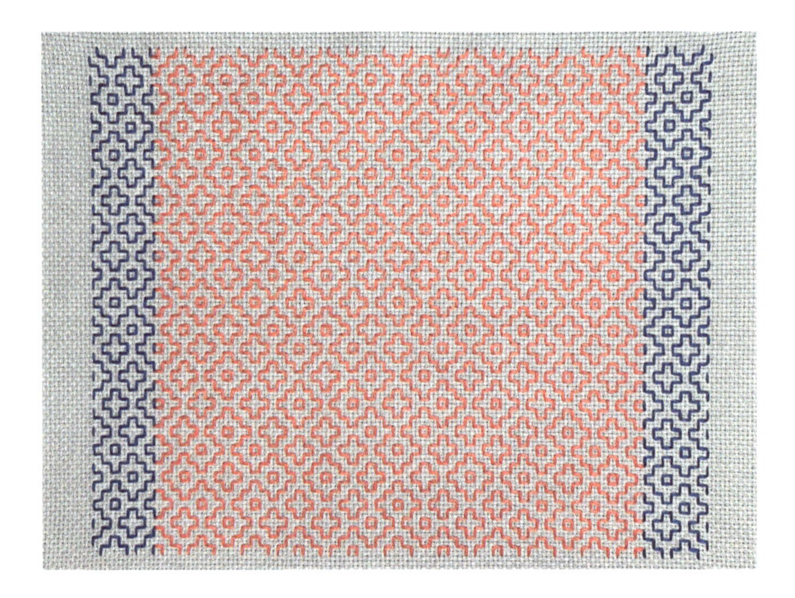

Figure 8.10 Generalised Pell persimmon polyomino pattern ($i = 4,\ k = 3$) in orange. Stitching and photograph by the author.

pair $P_{n-1}^{[j]}$ and $P_n^{[j]}$ one is even (2ℓ, say) and the other one is odd ($2m+1$, say). Hence the area of any of the generalised snowflakes is

$$4(\ell^2 + m^2 + m) + 1 \equiv 1 \bmod 4$$

as required for a hitomezashi loop.

The perimeter length, in terms of the generalised Fibonacci numbers, is $4F_{3k-1}^{[i]}$. We see from the values in Table 8.5 (and can prove in general by induction) that this total will be congruent to 4 modulo 8.

Once again, it is a conjecture that the largest motifs in the i-generalised Pell persimmon polyomino patterns are the i-generalised Fibonacci snowflakes for all types and orders.

You've probably realised by now that the flag labels in *Snowflake Bunting* refer to the type (first number) and the order (second number), with the line indicating the binary complement pattern (to see the self-dual nature). The orange stitching in Figure 8.10 shows the $i = 4,\ k = 3$ Pell persimmon polyomino pattern. There is an infinite number of these patterns—by both type and order—and one could spend a lifetime stitching or drawing them.

In preceding section(s) the construction placed the largest Pell persimmons in a square array, with the symmetry that we know from Chapter 6 is called p4mm. As we can see by comparing Coasters D and G from

Chapter 4 (see Figures 4.11 and 4.12), the smaller motifs can also appear traditionally in a half-drop or c2mm design, which is not self-dual. This next sampler is based on such an offset arrangement of the larger type 4, order 2 snowflakes. Again, a swirl of other snowflakes appear between, within, and on the reverse of these.

8.7 SAMPLER: SNOW AT CHRISTMAS

This piece makes a nice seasonal table runner or wall hanging. My piece has a rod pocket on the reverse, so that I can use it either way.
Required materials:

- Fabric, thread, needle and scissors; or grid paper, pencil, eraser and ruler.
- Backing fabric to make a table runner and/or a rod to make a wall hanging.

Planning: The longest dimension of the piece should be a multiple of 8 lines of stitching. The photo shows a piece which is 120 lines in one direction, 42 in the other.
Instructions: The design we will stitch is shown in Figure 8.11. The vertical lines of stitching are identical to those of bunting flag labelled

Figure 8.11 Hitomezashi piece: *Snow at Christmas.* Stitching and photograph by the author.

42, repeated (see Table 8.1). The states of the 42 horizontal rows are

$$0100110110110010\ 10010\ 01001\ 0100110110110010.$$

In terms of generalised Pell words, between the words $p_2^{[4]}$ and $\widetilde{p_2^{[4]}}$ we see $p_1^{[4]} = 101$ and $p_0^{[4]} = 01$, and their reversals and binary complements.

8.8 MORE TO EXPLORE

- It was stated without justification that 11 never appears in the Fibonacci words. Can you explain why? What about 000?
- Prove (by induction) that the Fibonacci numbers F_{3k+2} where $k = 0, 1, 2, \ldots$ are even. (Again, this was stated with no justification.)
- Prove (by induction) that the generalised Fibonacci numbers $F_{3k-1}^{[i]}$ for even i are odd.
- Stitch or draw some of the generalised Pell persimmon polyomino patterns, for higher values of i and k.
- Stitch or draw more half-drop patterns using the hitomezashi motifs defined in this chapter.
- Can you think of other ways to make some *Fractal Bunting*? There's a gallery of ideas to get you started at the Aperiodical website.

CHAPTER 9

Quasiperiodic patterns

BETWEEN randomness and the complete regularity which gives rise to symmetry, is the realm of the quasiperiodic. In this chapter, several well-known quasiperiodic sequences are used as stitching instructions for hitomezashi.

9.1 SAMPLER: QUASIMODO

The phrase *quasi modo* is Latin for 'in the manner of' and is used in that sense here (and not as a reference to the bell-ringing character from a Victor Hugo novel).

Required materials:

- Writing materials
- Fabric, thread, needle and scissors; or grid paper, pencil, eraser and ruler

Planning: For this sampler, you will first generate the stitching instructions. Begin with $w_0 = 0$ and apply this recurrence relation for the words

$$w_{n+1} = w_n\overline{w_n}. \tag{9.1}$$

The next two are: $w_1 = 01$, $w_2 = 0110$. The words double in length in each step, so that $|w_k| = 2^k$. Decide how large you wish your stitching area to be. It need not be square. Generate words to the length required. The piece in Figure 9.1 was stitched on a purchased cushion cover printed with a square grid on each side, the grid on one side having double the spacing of the grid on the reverse. On one side, 32 lines of stitching were worked in each direction, and on the other, 64.

DOI: 10.1201/9781003392354-9

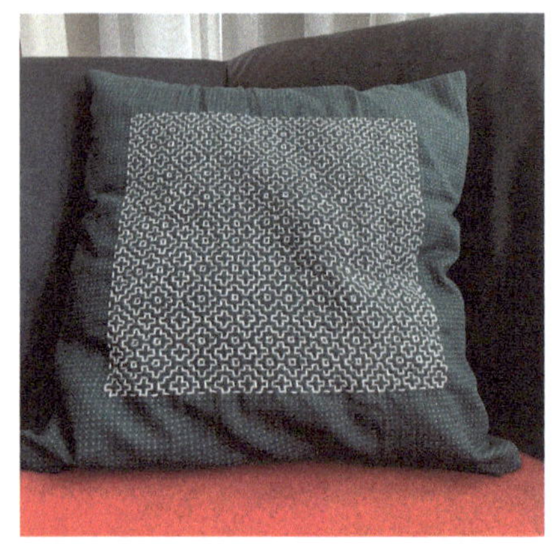

Figure 9.1 The design *Quasimodo* has been stitched at different scale on the two sides of a cushion cover. Stitching and photograph by the author.

Instructions: The same binary sequence or word is used for the vertical and horizontal lines of stitching. If you are stitching a rectangle, use more of the sequence in one direction than in the other.
Reflection: What do you notice? What do you wonder? Apart from the diagonal symmetry, what other features do you notice?

9.2 LONG WORDS

One way to create an arbitrarily long word (or string or sequence) is to repeat a finite word over and over again. This creates a *periodic* object, the kind we have used, particularly in Chapters 6 and 8, to create regular hitomezashi patterns with translational symmetry. The *period* is the length of the repeated unit. Asked what the 17th or 117th letter in the long word will be, there is a way to work it out. If the unit we're repeating has 8 digits, then the $17^{\text{th}}$is the same as the first (and the 117th is the same as the fifth).

For an arbitrarily long random string, such as the ones we looked at in Chapter 5, to determine the 17th or 117th digit we have to generate that many random numbers by our preferred method (dice rolling or more likely from an online random number generator). There are no short cuts.

The sequence of digits we generate using Equation (9.1) many, many times is a *quasiperiodic* sequence. Both random and quasiperiodic sequences are non-periodic. Within a quasiperiodic sequence, although there is no overall periodicity, each subpart can be found repeated again and again.

Using this one-dimensional quasiperiodic sequence to create our sampler *Quasimodo* has given rise to a pattern which has tantalising little patches which seem to have rotational or reflection symmetry locally. The pattern falls somewhere between uncompromising regularity and total randomness. We see motifs repeated, but there is no translational symmetry. That is, if we had two copies of our sampler, one on top of the other, we couldn't lift up the top one, move it without rotation, and put it down again as a perfect match. Certainly, we could match up some of the pattern, but the various elements do not have the same relative positions to each other when they recur. (There is also overall diagonal reflection symmetry which is unsurprising; we explicitly constructed it in when we chose identical stitching instructions horizontally and vertically.)

Mathematical interest in quasiperiodic sequences has been advanced by applications and explorations in theoretical computer science. Investigation of two-dimensional quasiperiodic patterns was fuelled by the discovery in the 1980s of non-periodic physical materials which nevertheless show signs of order in their diffraction patterns, thus shaking two centuries of accepted wisdom in materials science. These materials are called *quasicrystals*. We will come back to the influence of quasiperiodicity in generative art, but first we will examine the sequence generated from Equation 9.1 more closely.

9.3 THE THUE-MORSE SEQUENCE

The Thue-Morse sequence is named for two mathematicians: the Norwegian Axel Thue (1863–1922) and American Marston Morse (1892–1977). Morse was a distant cousin of Samuel Morse (from Chapter 7). There are various methods which give rise to the sequence, in addition to the recursive method we've already used.

Substitution: Using the morphism $0 \rightarrow 01$; $1 \rightarrow 10$ we obtain, beginning with 0:

$$0 \quad 01 \quad 0110 \quad 01101001 \quad 0110100110010110 \quad \ldots$$

The Thue-Morse sequence is a *morphic word*, which means as we make this substitution to generate longer words, the part we already have does not change. The word has a stable configuration, as far out as we could possibly want to follow it.

Figure 9.2 The motif on the right appears in Thue-Morse-inspired hitomezashi, but the igeta motif (on the left) cannot. Created by the author.

Parity: The nth letter of the word is the parity of the sum of the digits in the base two representation of n. For example, $17 = 2^4 + 2^1 = 10001_2$ and $117 = 1110101_2$. Summing the digits in each number, we get 2 and 5 respectively. Now interpreting these modulo two, the 17th digit of the Thue-Morse word is 0 and the 117th is 1. (Remember to start enumerating at 0.)

We see some interesting hitomezashi motifs in *Quasimodo*. The motif on the right in Figure 9.2 does not appear to be listed in any stitch dictionaries of traditional hitomezashi elements. It is a fusion of four persimmons, and seems to be a cousin of the traditional igeta motif, shown to its left. These motifs can both be found in the table of images of pseudo-squares given by Blondin Massé et al. (2010), as is the larger version (fusion of four double persimmons) which is also to be found on the diagonal symmetry axis in Figure 9.1.

However, the igeta motif itself will not occur anywhere in a pattern generated by Thue-Morse sequence. The Thue-Morse sequence is known to contain no cubes, that is digits (or sub-words) repeated three times in a row. The binary sequence for stitching igetazashi contains a quadruple, being 011110. The triple persimmon stitch also cannot appear: the stitching instruction contains the cube 101010. In fact, the motifs cannot be any larger than those we have seen in Figure 9.1; fusion of two or four ordinary or double persimmons.

The symmetries that we see in *Quasimodo* are also interesting. In a square with 2^k stitches, the symmetry depends on the parity of k. When k is even, the symmetry is D_4. When k is odd, we see only the construction symmetry across one diagonal, D_1. However, here the interesting relationship is between the front and back of the work. The odd k patterns are self-dual under a rotation through 180° about the centre of the square.

This might cause us to ask about another pattern we have seen which has D_4 symmetry but no translational symmetry. Can the infinite persimmon pattern (Figure 3.2) be called aperiodic? The answer to that is no. The definition of aperiodicity explicitly excludes non-periodic patterns which have infinitely large periodic sections. The 'infinite' persimmon is generated by imposing a local disruption on an otherwise periodic sequence.

9.4 QUASIPERIODIC ART

The motif on the right of Figure 9.2 is found in the digital artwork of Mark Dow (2007), reproduced by Campbell (2022). Campbell also points us as readers to the Thue-Morse artwork of Adam Ponting (2016) and weavable designs by Ahmed (2013), and uses the sequence in his own computer-generated artwork. There is a close relationship between the *automatic sequences* of theoretical computer science (generated by an automaton) and morphic words (generated by a substitution rule). Campbell uses the first term.

Several participants in the *Math Art Challenge* used the Thue-Morse sequence to generate hitomezashi drawings either drawing them by hand or programming them. In particular, Rodrigo Angelo tweeted the digital image shown in Figure 9.3, using black, white and grey to accentuate the connected regions of various sizes.

In weaving, the zeroes and ones become unders and overs, much as they do in hitomezashi. In the making of bobbin lace, threads are braided together, interacting in groups at regular intervals and then moving on, forming osculating paths on a grid. By making the spacing between the interactions to be either long or short, Irvine, Biedl and Kaplan (2020) used words on the two-letter alphabet $\{L, S\}$ to design quasiperiodic bobbin lace. In particular relevance to this chapter, Veronika Irvine made a lace ground determined by the Fibonacci word. This lace and the hitomezashi design shown in Figure 9.4 provide different two-dimensional visualisations of the same quasiperiodic sequence, through different fibre arts. *Fibonacci* is intended as a companion piece to that in Figure 6.2.

The alphabet $\{L, S\}$ plays a key role in a model for quasicrystals called the *labyrinth tiling*. This tiling is generated using the octa- or octo-nacci sequence, and approximates the octagonal quasiperiodic tiling

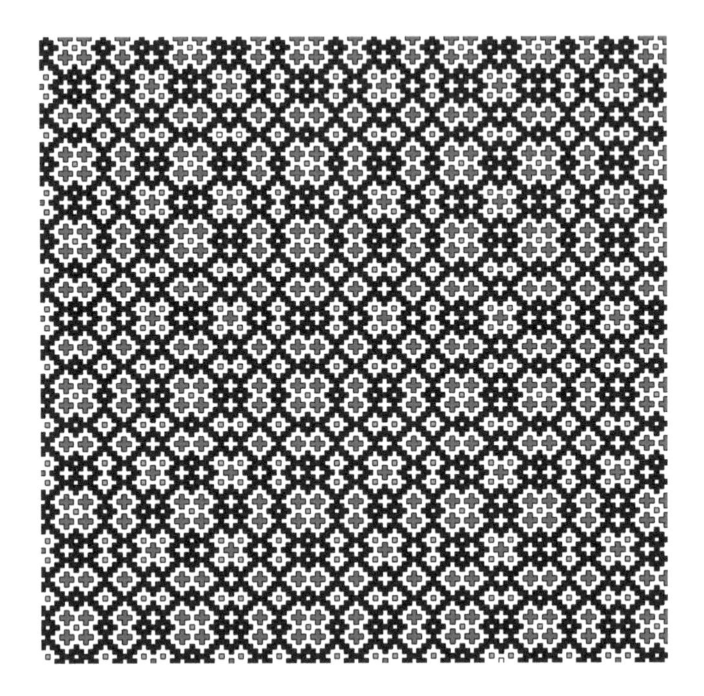

Figure 9.3 Thue-Morse hitomezashi on a square grid of size $130 \approx 2^7$. Created by Rodrigo Angelo (Brazil) and used with his kind permission.

[Sire et al., 1989]. The sequence name refers to this tiling and to similarity with the Fibonacci word. The recurrence relation for the octanacci sequence strongly resembles our Equation (8.4), being

$$S_{n+1} = S_n S_{n-1} S_n, \quad n \geqslant 1 \tag{9.2}$$

with initial sequences $S_0 = L$ and $S_1 = S$. The length of these sequences is given by a generalisation of the Pell numbers (though not that of Equation (8.5)). $|S_n|$ is found using Equation (8.2) with initial numbers $P_0 = P_1 = 1$. The first few are 1, 1, 3, 7, 17, 41.

The hitomezashi piece *Octanacci* shown in Figure 9.5 has been stitched using the sequence generated using Equation (9.2) with $S_0 = 0$ and $S_1 = 1$. Like the Thue-Morse sequence, this sequence can also be generated using a morphism:

$$0 \to 010 \quad 1 \to 0.$$

Quasiperiodic art has reassuring familiarity, as we spot the repetition of motifs, along with an element of surprise, as the local environment of

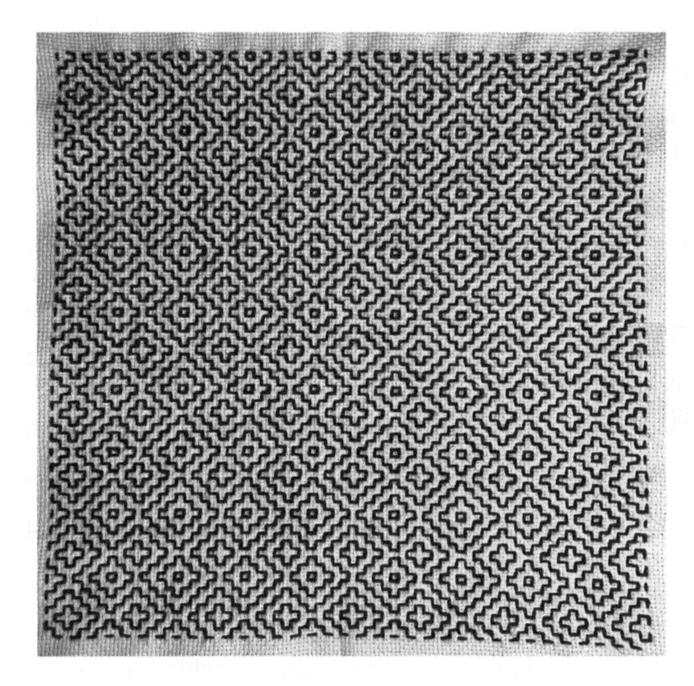

Figure 9.4 Hitomezashi piece: *Fibonacci* made using the Fibonacci word for stitching instructions in both directions. Stitching and photograph by the author.

a motif is different at each repeat. The visual hitomezashi representation arguably makes quasiperiodicity more apparent to our senses than inspection of a string of digits. Symmetry and fractals have inspired mathematical fibre artists, but it seems that thus far collectively we've only scratched the surface of the potential of using quasiperiodicity in our designs.

> "... it is precisely the non-periodicity and non-repetitious characteristics of [automatic] sequences ... juxtaposed with the systematic ways in which these sequences are generated, that form a basis of so much about the aesthetic appeal of [related] works of art ..."
>
> J. M. Campbell
> [Campbell, 2022, p. 289]

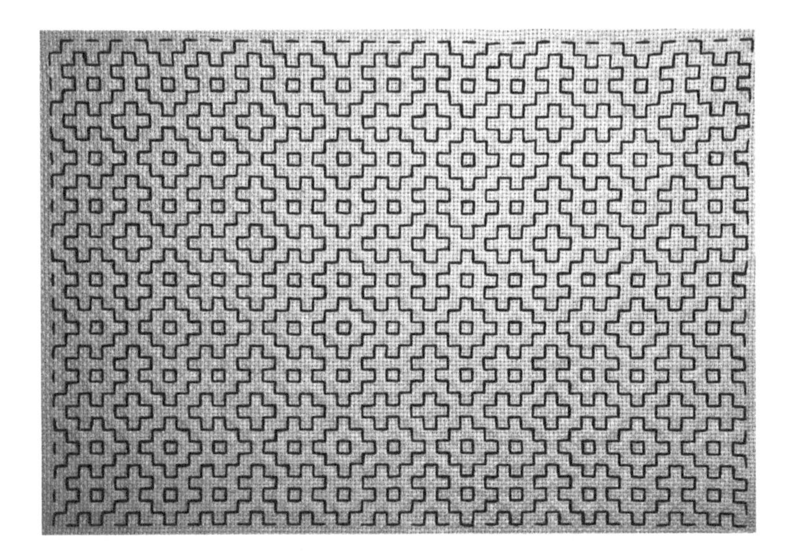

Figure 9.5 Hitomezashi piece *Octanacci* made using the octanacci sequence for stitching instructions in both directions. The diagonal symmetry axis runs from the upper left corner. Stitching and photograph by the author.

9.5 MORE TO EXPLORE

- Generate the octanacci sequence.
- Stitch or draw hitomezashi with instructions obtained from other quasiperiodic sequences. You can find many such sequences in the Online Encyclopaedia of Integer Sequences [OEIS Foundation Inc., 2024], such as the Rudin-Shapiro sequence (A020987) and the paperfolding or dragon curve sequence (A014707).
- Explore the effect of using one quasiperiodic sequence for the vertical stitching instructions and a different one for the horizontal.
- What might quasiperiodic knitting look like? It would fall somewhere between random lace or cables (see Figure 5.3), and traditional lace or cables. If you're a knitter, you might like to explore this idea.
- Hitomezashi designs with the regions coloured have the appearance of mosaic or patchwork. If these interest you, you could make a quasiperiodic piece based on Angelo's design in Figure 9.3.

CHAPTER 10

Corners

HITOMEZASHI is created using horizontal and vertical stitches, which interact with each other forming right-angled corners. How corner percolation, paths on the square lattice, and the rounded corners depicted on Truchet tiles connect to hitomezashi—and to each other— is explored in this chapter.

10.1 SAMPLER: STRATA

Required materials:

- Fabric, thread, needle and scissors; or grid paper, pencil, eraser and ruler.

Planning: This piece lends itself to being worked on a rectangle in landscape orientation. Worked in blue and white it resembles stratocumulus clouds; worked in earth tones, the layers of sedimentary rock formations. Choose the size you want your finished piece to be, and determine how many horizontal and vertical lines of stitching you will work.

Instructions: The horizontal rows alternate in state 10101010.... The vertical lines alternate between short sequences in which the lines are all of the same state or in which states alternate like the horizontal rows. (See Figure 10.1.)

Reflection: In hitomezashi we see only loops (polygons) or paths that extend (theoretically to infinity) in both directions, which we describe as bi-infinite. That is, we never see a spiral, or a loop with a tail like a comet or a tadpole. The designs on some of the coasters in Chapter 3 —dan tsunagi and jōkaku—feature no loops, and it's obvious that we could continue them indefinitely as bi-infinite paths. Can you see how in *Strata*

DOI: 10.1201/9781003392354-10

Figure 10.1 *Strata* has lines of stitching, but no loops. Stitching and photograph by the author.

these two stitch patterns have been combined to create the layered effect which gives the sampler its name?

Look back at the random hitomezashi created in Chapter 5, particularly *Ever so Airy a Thread*, and the quasiperiodic pieces from Chapter 9. Do the stitches form any continuous paths which travel however indirectly between opposite edges of the piece? (Since everyone's execution of a random sampler will be unique, there may not be any in yours. In the piece shown as Figure 5.8, there is one such line.)

10.2 LOOPS AND WALKS ON THE SQUARE LATTICE

Because of my research background in statistical mechanics, in particular models on the square lattice, the patterns formed by hitomezashi stitches seemed familiar when I first encountered them. They reminded me of loop diagrams derived from vertex models. Loop-free designs resemble constrained self-avoiding walks. We've already borrowed the term fully packed from statistical mechanics: a configuration is said to be fully packed if every vertex of the underlying grid is visited by a loop. That is, none of the internal vertices have degree less than two. Perhaps the model closest to hitomezashi in its permitted configurations is dilute oriented loops [Vernier et al., 2016].

However, there is no actual 'physics' in hitomezashi, no attractive or repulsive force causing particular configurations to be energetically favoured. The kind of questions we might ask about its loops and paths fall under the umbrella, rather, of *enumerative combinatorics.*

A well-established connection exists between the six-vertex model, fully packed loops and *alternating-sign matrices* (ASMs) [de Gier, 2009]. Every entry in an ASM is one of $0, -1, +1$, and in each row and column entries add to 1 and the non-zero entries alternate in sign. A well-known subclass of ASMs are *permutation matrices.* In a permutation matrix, there is only one non-zero entry in each row and column, so it is necessarily 1. That is, the rows of a permutation matrix are a permutation of the rows of the identity matrix, and when another matrix is multiplied by a permutation matrix, the outcome is that its rows are permuted. By way of example

$$\begin{bmatrix} 0 & 1 & 0 & 0 \\ 0 & 0 & 0 & 1 \\ 0 & 0 & 1 & 0 \\ 1 & 0 & 0 & 0 \end{bmatrix} \begin{bmatrix} 1 \\ 2 \\ 3 \\ 4 \end{bmatrix} = \begin{bmatrix} 2 \\ 4 \\ 3 \\ 1 \end{bmatrix}.$$

Labbé (2010) produced a remarkable image showing Fibonacci snowflakes emerging in the fully packed loop diagram corresponding to a particular 32×32 permutation matrix. I have stitched this as a piece of hitomezashi, shown in Figure 10.2. At each position corresponding to a non-zero matrix entry, a 'defect' (to borrow another term from statistical physics) arises in the alternating state of the stitches, marked with a bead. In the correspondence between ASM entries and loop configurations, the entry 1 causes the loop *not* to turn at that point. Not only does the bead indicate where the defect occurs, quite practically the thread has been anchored by the bead as it traverses twice its usual distance. Even more notably, this design has a symmetry that we cannot render in defect-free hitomezashi. The full symmetry of the piece is C_4.

A loop-free connected walk on a lattice which visits every vertex is called a *Hamiltonian walk.* Such a walk was employed by photographer Daniel Crook in the *Portrait* series in his exhibition *Remapping.* Each work consists of multiple images of the subject across twenty minutes, captured by a camera mounted on a robot following a mathematically accurate Hamiltonian walk [Crook, 2012].

Figure 10.2 Fibonacci snowflake design in hitomezashi stitching based on a permutation matrix. Stitching and photograph by the author.

The four 'corners' of Figure 10.3, a degenerate case of the six-vertex model, feature in a problem termed corner percolation and posed by Hungarian mathematician Bálint Tóth.

10.3 CORNER PERCOLATION

When we make coffee by percolation, hot water finds paths (percolates) through tiny gaps in a thick layer of ground-up coffee beans. The water being hot, chemical compounds are dissolved into it from the fresh grounds as it percolates; the temperature of the water and the grind of the coffee beans affect the flavour achieved. While another Hungarian mathematician Alfréd Rényi famously quipped that 'a mathematician

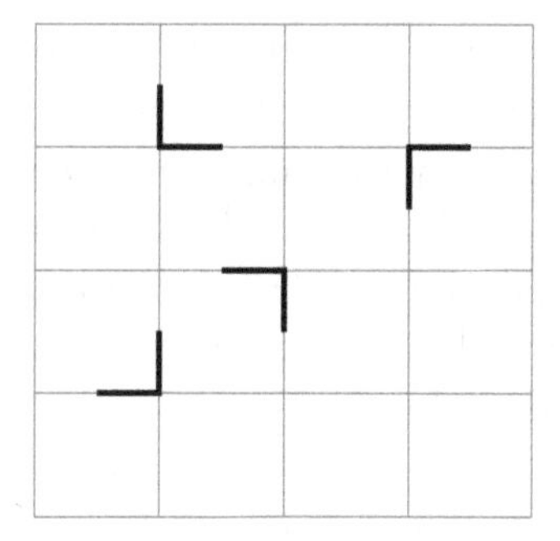

Figure 10.3 The four possible corner configurations formed by fully packed hitomezashi. Created by the author.

is a device for turning coffee into theorems', *percolation theory* is more about the theory and disappointingly not at all about coffee.

Mathematical percolation considers the nature of the networks formed when bonds (or occupied sites) are placed randomly onto a theoretically infinite set of vertices. As the probability of the presence of bonds—which we can think of as open channels through which a liquid could flow—is varied, the connectivity between the vertices changes. At some particular value of the probability, there can be a phase transition, from a state of overall unconnectedness, to one in which there is a so-called giant connected component.

Pete (2008) studied the corner percolation model, in which one of the four kinds of corners (see Figure 10.3) must be created at each vertex when bonds are added. Applying this condition consistently has non-local effects. This is a form of dependent or constrained percolation. Placing bonds for corner percolation is identical to placing stitches in hitomezashi. That is, each whole row or column of bonds can be in only one of two states, entirely determined by the first one placed in that row or column; the other bonds therein cannot later be placed randomly. We met some of Pete's findings about loops in Chapter 3. Pete further showed that if the two states for the lines of stitching or bonds are equally probable, there is no giant connected component. Rather, as the grid size becomes infinite, there are infinitely many closed loops. The intriguing square root of 17 in the title of his paper arises in the behaviour of the expected value of the length of loops with diameter n—roughly what we've called width— as n gets large. It behaves like n^δ where $\delta = \frac{1+\sqrt{17}}{4}$.

More recently, Marchand et al. (2022) have considered corner percolation by treating each bond (or hitomezashi stitch) as a step in a two-choice, must-turn walk with long-term memory taken on the grid. Assigning probabilities $q \neq \frac{1}{2}$ and $p \neq \frac{1}{2}$ to steps which move north and west respectively (so that steps to the south have probability $1-q$ and steps east, $1-p$) they found different behaviour when there are preferred directions. For this model, the configuration in the large-lattice limit is not one of loops, but is comprised of an infinite number of bi-infinite paths. The slope of the paths approaches $\frac{2q-1}{1-2p}$.

10.4 TRUCHET TILES

In the early 1700s, the French priest and engineer Sébastien Truchet noted, among some building materials, decorated ceramic square tiles which could be placed in four orientations when used to tile a floor (Figure 10.4). The essence of his discussion appears translated from French, with reproductions of the original plates and tables of illustrations, within a longer paper by Smith and Boucher (1987). Truchet considered the number of ways in which two tiles could be placed together and illustrated 30 larger designs with various symmetries created by combining the tiles; the mathematics of combinations was at that time a nascent area of research.

Truchet reported that he had consulted the mathematics literature on combinations and books on civil architecture to determine the novelty of his observations, but found nothing similar. Could he have been looking in the wrong place? Numerous designs involving the simple

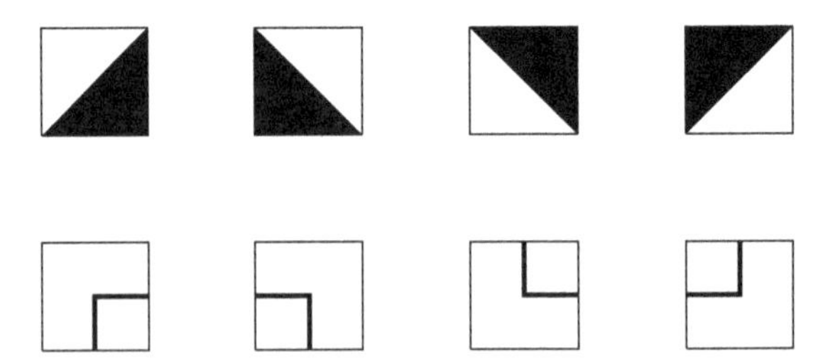

Figure 10.4 The top row shows the four orientations of a diagonally half-coloured square known as Truchet tiles. In the bottom row the four configurations possible on the dual grid in hitomezashi are displayed, directly beneath the Truchet tile to which they can be mapped. Created by the author.

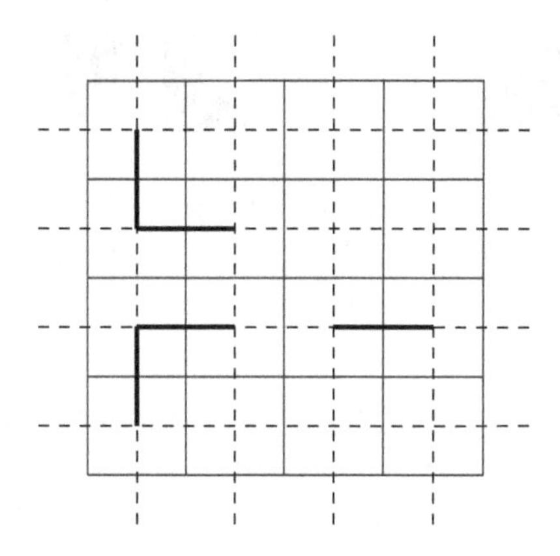

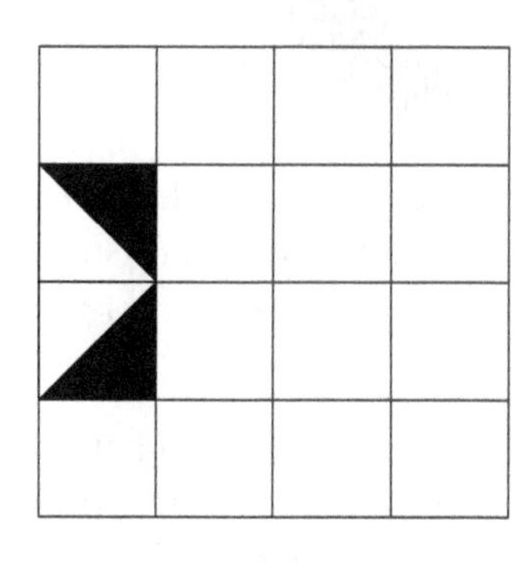

Figure 10.5 On the left, some hitomezashi stitches on the square grid, showing the dual grid; on the right, the corresponding Truchet tiles on the dual grid. Created by the author.

half-square triangle are known in patchwork, but being chiefly a domestic art created from perishable material, the record of its history is patchy (to use a rather unsubtle pun). Janniere (1994) argues that patchwork was practised in France in the relevant time period.

Instead of thinking of the four corners in Figure 10.3 as occupying vertices of the square grid, we can think of them as living at the centre of the faces of the dual grid (to which we were introduced in Chapter 4). In Figure 10.5 the dashed black lines indicate a square grid. Some hitomezashi stitches—thick black line segments—have been placed on this grid. The blue lines mark the dual grid. Half of each hitomezashi stitch lies in one face of the dual grid, and the other half lies in an adjacent face. Once all the hitomezashi stitches have been completed, each face of the dual grid will contain one right-angled corner. We can map a vertex configuration on the dual grid to a tiling with Truchet tiles, by colouring the faces as shown in Figure 10.4.

Because each half-stitch comprising a corner must be completed in the adjacent face, when the placement of Truchet tiles is derived from hitomezashi it is tightly constrained. Black (or white) triangles can meet only to form a parallelogram or a larger triangle (see Figure 10.6). Of the 30 designs illustrated by Truchet, in only 5 of them is this rule obeyed.

One of these—which is image Y in Truchet's plate 4, and has been recreated in part as Figure 10.7—corresponds to the stitched pattern on coaster G in Figure 4.12. The self-duality of this design is visually apparent in the tiling, the white regions indicating stitches on the back

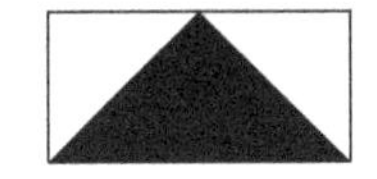

Figure 10.6 Truchet tiles permitted to be adjacent when mapped to hitomezashi (left hand and centre image). The right hand arrangement does not correspond to hitomezashi. Created by the author.

of the work. The Truchet tiling corresponding to the hitomezashi dual can be obtained by swapping black and white.

Towards the end of his paper, Smith introduces a different decoration of a square tile, which has two orientations. These have also come to be called Truchet tiles; arguably they are better known than the original tiles which bear his name. More properly and for clarity they are referred to as Smith or Smith-Truchet tiles. The tile in its two orientations is shown in Figure 10.8. When the tiles are placed together, the quarter circles always connect to form meandering curved paths and loops, permitting a two-colouring of the outlined regions. From artistic images to the bathrooms of the National Museum of Mathematics (MoMath), Smith-Truchet tiles have become a favourite in mathematical art and outreach activities (see, for example, [Lawrence, 2018] and [Bosch, 2012]). The mathematics of the tiles is widely accessible, as Smith observed:

Figure 10.7 Part of an design created by Truchet which corresponds to the traditional hitomezashi design of crosses and squares. Created by the author.

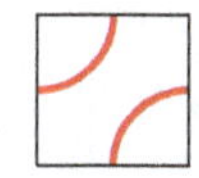 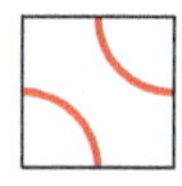

Figure 10.8 Two orientations of the tile introduced by Cyril Stanley Smith. Created by the author.

> "...the tiles suggest a departure from customary emphasis on either geometry or analysis via trigonometric functions and reciprocals and invite the adoption of the simpler topological approach... It involves neither infinitesimals nor negative or irrational numbers—so essential in many calculations..."
>
> Cyril Stanley Smith
> [Smith and Boucher, 1987, p. 378]

We can substitute quarter circles sharing endpoints for the right-angled corners of hitomezashi, as shown in Figure 10.9, but this does not give whole Smith-Truchet tiles. David Reimann (2023) has pointed out where the missing half of the decoration can be found—it's on the reverse, part of the dual pattern! In Figure 10.10 we can trace his argument. The image on the left shows quarter circles (red) superposed on a hitomezashi design (white) and the dual grid (yellow). In the centre

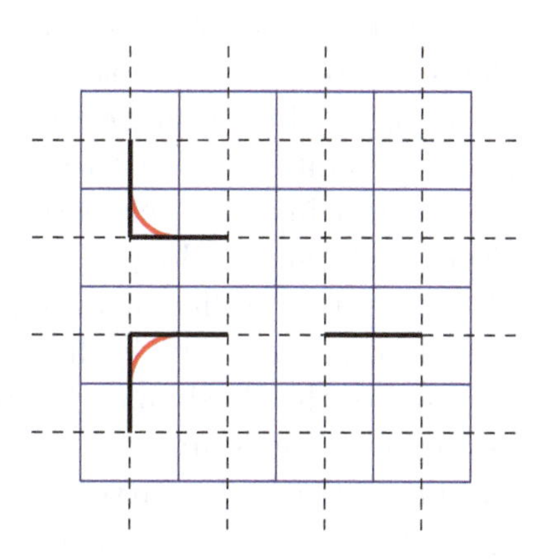

Figure 10.9 Quarter circle arcs corresponding to hitomezashi corners. Created by the author.

Figure 10.10 The steps in mapping a hitomezashi design to a Smith-Truchet tiling. Created by David A. Reimann (USA) and used with his kind permission.

image, the arcs corresponding to the dual design have been added in pink. Finally, the image on the right shows only the Smith-Truchet curves.

There are many more possible arrangements of Truchet or Smith-Truchet tiles than there are hitomezashi designs. On a rectangle with m rows and n columns, there are 2^{m+n} possible random hitomezashi designs. But if the tiles are placed randomly, there are 4^{mn} possible Truchet tilings, and 2^{mn} possible arrangements of the Smith-Truchet tiles. Hitomezashi-compatible Smith-Truchet tilings must have an even number of each tiles of each orientation in any 2×2 block [Reimann, 2023].

One particular arrangement of Smith-Truchet tiles which corresponds to the Fibonacci snowflakes that we stitched in Chapter 8 features in digital artwork by Hooper (2019). Being (not surprisingly) unaware of the dual of hitomezashi stitching, Hooper made a two-to-one mapping of corner percolation vertices to Smith-Truchet tiles. This Smith-Truchet-Fibonacci snowflake has special invariant properties in the context of a particular renormalisation of tilings [Hooper, 2013]. The fascinating connection between what they term different avatars of linear Truchet tilings is the subject of current research by Gradit and van Dongen, and appears relevant to the proof of the Persimmon-Snowflake conjecture.

Squares with a inset quarter circle are well known in the folk-art-turned-fine-art of patchwork, though the curve makes them more difficult to sew than a design with straight seams. The fraction of the square occupied by the quarter circle can be varied. Sewn together in different orientations, they give designs known as Solomon's Puzzle, The Mill Wheel or Drunkard's Path [McKim, 1962]. There are also patchwork designs which use the Smith-Truchet tile (and which predate 1987 by decades). Mary Shepherd (2018) has described various types of connected paths that can be created by careful placement of these so-called Snake's Trail squares: Snake Paths, Snake Trails, Racetrack and Snake in a Hollow Maze. Knowledge of how to create a maze or a continuous path which visits every square, an existence proof of the solution to a mathematical puzzle, could be held as a closely-guarded secret by a quilter. As Shepherd shows, these solutions can be deduced using graph theory or by systematic re-orientation of squares. It is only Snake Paths and Snake Trails which correspond, via Smith-Truchet tiles, to permitted hitomezashi designs.

10.5 MORE TO EXPLORE

- Prove (by contradiction) that a hitomezashi pattern on a rectangle cannot have both a line of stitching that crosses from top to bottom and from left to right.
- Use permutation matrices to create hitomezashi with defects, as described in Section 10.2.
- Draw some of the Truchet tile designs that correspond to hitomezashi stitch patterns from earlier samplers.
- If you are a patchwork enthusiast—or novice—make a sampler either using the half square triangle (Truchet's original tiles) or Drunkard's Path or Snake Trail squares (Smith-Truchet tiles).

CHAPTER 11

Off the grid

HITOMEZASHI is taken off the plane in this chapter, to decorate the surface of three dimensional fabric and paper objects. It is also taken off the monochrome square grid, by adding diagonal stitches and colour, by varying stitch length and by stitching on a triangular grid.

11.1 SAMPLER: HITOMEZASHI OCTAHEDRON

In this sampler, you will draw a hitomezashi design onto six squares of grid paper, before folding them and weaving them into a piece of modular origami called Neale's skeletal octahedron.

Required materials:

- Grid paper, scissors, pencil or erasable pen, eraser and ruler.
- Optional: six square sheets of paper, two each of three colours.

Planning: For this sampler, you will need to cut six squares of grid paper. The design used in Figure 11.1 is 32 grid squares wide, on grid paper with a 5 mm grid spacing, so the paper has been cut to 16 $\times$16 cm.

An optional preparatory step is to make Neale's octahedron from coloured paper, as in Figure 11.2, in order to understand its modular construction. Use the instructions which are provided on the website Origami Heaven [Mitchell, 2016]. It can be helpful to use paperclips to hold the pieces temporarily until the sixth module locks them all together. Once you've combined the modules, you may like to mark with pencil the triangular regions that are visible in the assembled piece. Of course, to see which parts are not marked with pencil, you have to be brave enough to take the octahedron apart again!

DOI: 10.1201/9781003392354-11

Figure 11.1 *Hitomezashi octahedron*: modular origami with hitomezashi decoration. Folding and photograph by the author.

Figure 11.2 Neale's skeletal octahedron made from two sheets of yellow, two sheets of grey and two sheets of recycled multi-colour Japanese wrapping paper. Folding and photograph by the author.

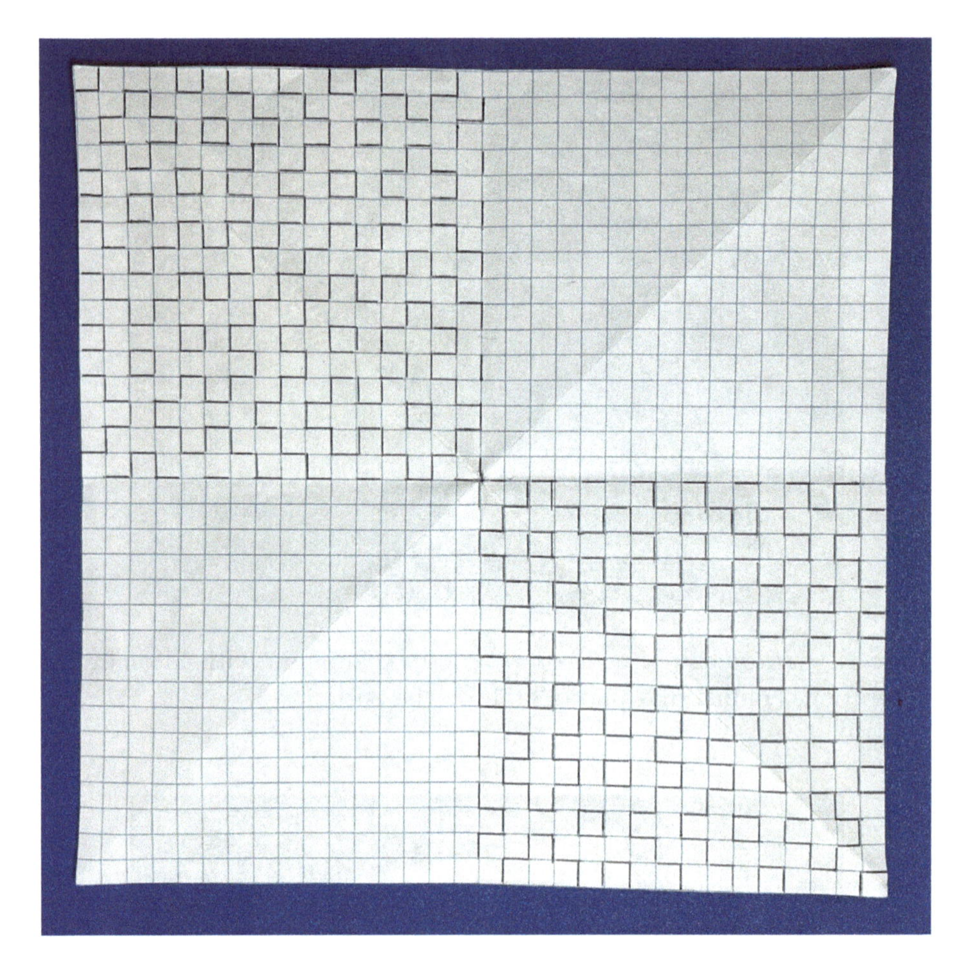

Figure 11.3 Drawing and creases for one module of *Hitomezashi octahedron*. Drawing and photograph by the author.

Instructions: It is only necessary to mark stitches on the two quadrants of each square that will be visible in the final piece (see Figure 11.3). The stitching (drawing) instructions are: 1001110101110010. This should first be done on one quadrant of a square, 16×16 grid squares. The paper is then rotated through 180° before the drawing is made on the second quadrant. (The centre vertex has degree four, but when assembled two stitches will fall into one octant and the other two into another.)

Draw the hitomezashi pattern onto each of six squares cut from grid paper. To mark up 12 quadrants with no errors is something of a challenge, and you may like to use an erasable pen. Then fold the squares into origami modules, and finally assemble them into the skeletal octahedron. (Use the instructions which are provided on the website Origami Heaven [Mitchell, 2016].)

Reflection: How is the symmetry of the drawn pattern important to the final object? Is the hitomezashi design correct, as it wraps from one plane to the next along the right-angled boundaries? What features of the drawing instructions ensure this?

11.2 ORIGAMI OCTAHEDRON

I came across Neale's skeletal octahedron on Day 42 of Annie Perkins' *Math Art Challenge* in 2020, making my first octahedron out of odd pieces of grid paper which were lying around in abundance because I'd been working on hitomezashi designs. I soon realised that I could draw hitomezashi onto the paper before I folded it, thus taking hitomezashi off the plane. I experimented with ways to keep the alternating pattern correct as the various triangular faces of the model meet. The centre of one square coincides with the corners of those it is interwoven with in the assembled model. The diagonals of the squares form the edges of the octahedron.

Neale's octahedron is a *nolid*, a solid with no volume, the origami forming its skeleton. The edges and vertices of the octahedron are created, but not its external faces. Compare it to the octahedron vase in Figure 11.1. The triangles that comprise the skeleton meet at right angles, and the model suggests the Cartesian axes in three dimensions. The history of this modular construction has been traced to the 1960s [Mitchell, 2024]; it is not a traditional Japanese origami model.

The module from which it is assembled has a much longer history in Japanese origami. It has come to be called (in English) the balloon or water bomb base. I folded my first water bomb base as a child in the 1970s, in front of a black-and-white television, following along with Robin Harbin (1908–1978), magician and origami evangelist. Thinking about the features that a hitomezashi pattern should have in order for it to interact nicely with the folds and angles of an origami model aligns with our theme of hitomezashi mathematics, and is something more to explore. We can't pursue the very much larger topic of the mathematics—and art—of origami in general, but there are many books and websites devoted to this topic if it piques your interest.

11.3 SAMPLER: OTEDAMA PINCUSHION

The first sampler in this chapter was made with paper and drawn lines. This sampler can only be made by sewing.

Required materials:

- Fabric, thread, needle, and scissors.
- Polyester filling or dried beans or stuffing pellets.

Instructions: Making the otedama starts simply enough, with four rectangles, which must be exactly twice as long as they are wide. Stitch the outline in double running stitch. It is much easier to stitch the required small pieces as part of a single larger piece of cloth, and then to cut them out. Figure 11.4 shows four such rectangles, which measure 2 inches by 4 inches, aida cloth thread count being specified per inch.

You can decorate the rectangles with any hitomezashi design of your choice. To use your otedama to illustrate interesting symmetry in three dimensions without unnecessary distraction, a stitching design which has D_2 symmetry on the strip is recommended; that is, symmetry across both vertical and horizontal axes. I have chosen kawari kuchizashi, the variation giving almost the effect of an optical illusion. The long lines have been stitched as 110011110011 and the short lines are all in the state labelled 1.

Figure 11.4 Pieces required for *Otedama pincushion* are more easily stitched before being cut out. Stitching and photograph by the author.

Figure 11.5 Assembly instructions for *Otedama pincushion*. Stitching and photograph by the author.

Assembling the otedama is a bit fiddlier. In Figure 11.5 one end of each rectangle has been attached to the side of another, in the pinwheel form shown. The coloured threads in the seam allowance indicate the parts of the rectangles which are next to be attached to one another. I used hand sewing. By stitching the double running stitch lines together, working back and forth between them, the piece is assembled right-side out (rather than inside out as it would be if the sewing was done with a machine). The final section of the seam is stitched once the piece has been lightly stuffed. Use polyester filling to make a pincushion. For otedama to play with, use dried beans or stuffing pellets. The finished object is shown in Figure 11.6.

Figure 11.6 Hitomezashi piece: *Otedama pincushion*. Stitching and photograph by the author.

11.4 AXIAL POINT SYMMETRY: BISCORNU AND OTEDAMA

An unadorned biscornu is shown in Figure 11.7. It is made from two squares, offset and sewn together with the midpoint of each edge of one square matched to a corner of the other. Once lightly stuffed with filling, it has an intriguing many-cornered shape to which its name refers; 'biscornu' means oddity in French. Biscornu became extremely popular with embroiderers and quilters in the early part of this century, but its origins could not be traced to any particular sewing tradition.

The symmetry group of a biscornu is that of a *square antiprism*. At the time that I created a complete symmetry sampler, by decorating biscornu with hitomezashi to represent the subgroups of its overall symmetry group, I was unaware of the centuries-old Japanese juggling game, *otedama*. It is played with drawstring bags containing dried beans, or the intriguingly constructed closed bags also called otedama, or *ojami* in some parts of Japan. A game like juggling or jacks, it was chiefly a girls' pastime which began to wane after the second world war [Henderson, 1991]. Some attribute this decline to the need for the beans with which the little bags were filled to be used as food, and others to the common decline of traditional home-made toys when mass-produced

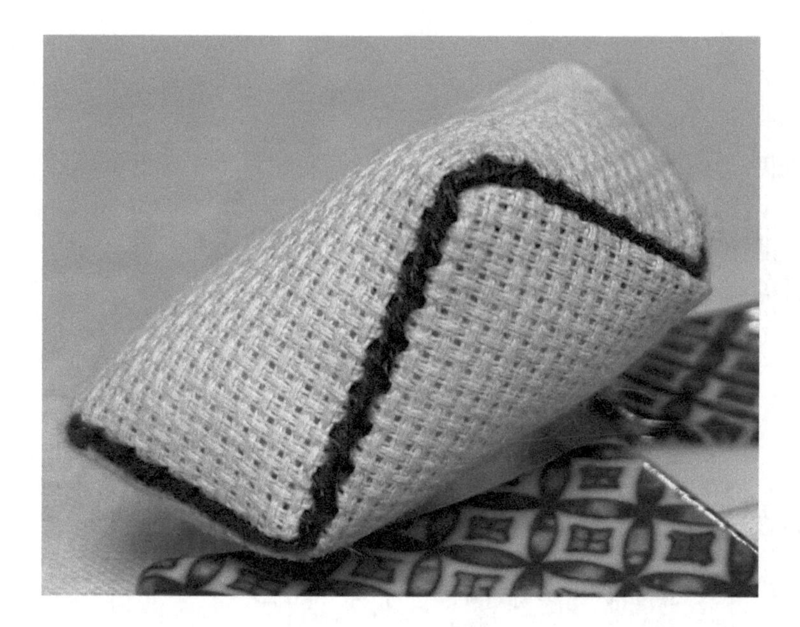

Figure 11.7 A biscornu, undecorated apart from its seam. Stitching and photograph by the author.

options became available. We won't sidetrack into the mathematics of juggling, but we can explore some more mathematics with our otedama.

In Chapter 6, we considered the symmetry of objects confined to the plane under various transformations which preserve length and angle. We could translate, rotate about a point, reflect across a line, and combine reflection and translation as a glide-reflection. Square antiprisms live in three dimensions; they are invariant under the transformations that belong to the axial point group D_{4d} (in Schoenflies notation).

Sixteen transformations form the elements of D_{4d} and there are 11 subgroups. Imagine an axis running vertically through the centre of both faces of a biscornu. The whole object including its seam is symmetric under an eight-fold roto-reflection. The roto-reflection rotates the object through 45° about this axis and reflects across a plane which is perpendicular to the axis and lies between top and bottom square faces. Like a glide-reflection in two dimensions, a roto-reflection in three dimensions is a single transformation. And, like any reflection, we can't actually physically perform it on an object. The biscornu or square antiprism is also symmetric under a two-fold rotation about a line in the plane described above; this line makes an angle which is one quarter of a right angle with the symmetry axes of the square faces.

Figure 11.8 Hitomezashi piece: *(Ir)regularity* (2021). Stitching and photograph by the author.

The seams of an otedama, on the other hand, can be seen by comparing Figures 11.6 and 11.7 to follow quite a different path from the single seam of the biscornu. Even without decoration an otedama does not have the full symmetry of the square antiprism. We cannot decorate multiple otedama to create a complete symmetry sampler of D_{4d}, as I did with 11 biscornu in the artwork *(Ir)regularity* shown in Figure 11.8.

An otedama is unchanged by the transformations belonging to the axial-point group C_{4v}, a subgroup of D_{4d}. A bird's-eye view of key features of an otedama is given in schematic form in Figure 11.9 (a). The seams which run diagonally from the vertices of the upper surface to those of the lower surface are indicated. The object has four-fold rotational symmetry about the vertical axis, indicated by a filled black dot. The red dashed line indicates the other rotation axis of the object, which lies in a plane midway between the upper and lower square surfaces. In Figure 11.9 (b), it is shown how this two-fold non-axial rotation changes the labelling order of the vertices from anti-clockwise to clockwise, and interchanges the upper and lower surfaces of the otedama.

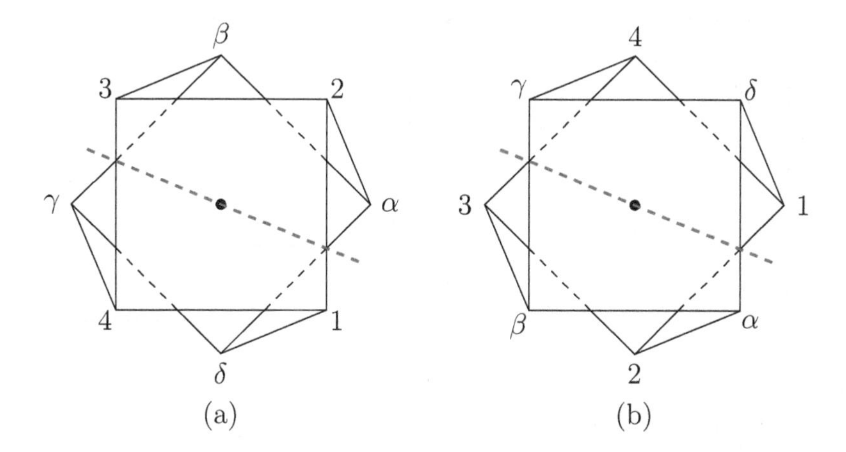

Figure 11.9 (a) Schematic diagram of an otedama with its axis and non-axial rotation axis. (b) The result of rotating through 180° about the red axis. Created by the author.

There is an inbuilt chirality to an otedama's construction, a twist which we can see even as we lay out the pieces for sewing (Figure 11.5). But this has exciting potential. To create biscornu for some of the cyclic subgroups, I resorted to hitomezashi stitching in which the vertices did not all have order two [Seaton, 2021], as I had to in Chapter 6 for some symmetries. An otedama is a decorated antiprism, with rounded edges. Each of the four rectangles used to construct it contributes one quarter of the upper square face and one quarter of the off-set lower square face, while the seams endow chirality which does not need to be provided by the hitomezashi with which we decorate it.

11.5 HITOMEZASHI ON THE TORUS

If a topologist were to look at a biscornu or odetama, they would see something that in their eyes is equivalent to a sphere. In fact, use too much filling and you will see that too. There is no hole in a sphere, unlike a *torus*, which to a topologist is equivalent to a doughnut or a cup with one handle. Loosely speaking, the number of holes in a surface is called its *genus* in topology. A sphere has genus zero, a torus has genus 1, and the sauce jug shown in Figure 7.2 has genus 2, the handle and the spout providing the two holes. One key difference between surfaces of different genus is the kind of loops that can be drawn on them. All loops on a

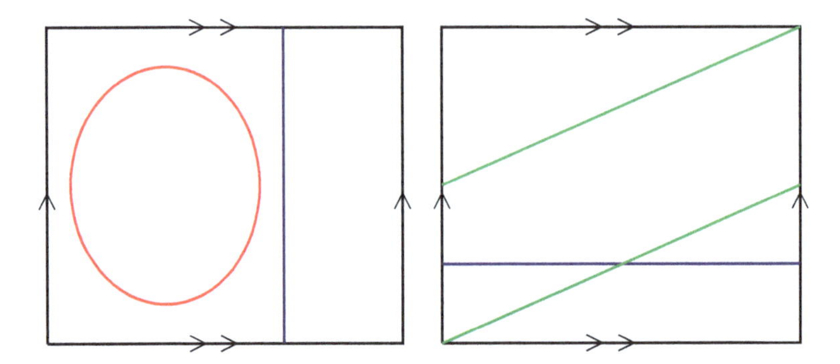

Figure 11.10 Loops marked in colour on two tori; arrows indicate identification of edges. Created by the author.

sphere are contractible. We can imagine shrinking a *contractible loop* on the surface to a single point.

In Figure 11.10 the rectangles each represent a torus. The arrow markings show how the edges of the rectangle should be glued together (or identified) to form the torus. A contractible loop is indicated in red, and loops which wrap around the body of the torus, and around the hole, are each drawn in blue (one on each torus). A loop can wrap both the body and the hole of the torus, and more than once. The loop in green in the right-hand image wraps twice. The loops indicated in blue and green are not contractible—the hole stops us from shrinking them to a point. Any two loops that wrap the hole and the body the same number of times can be contracted onto each other; they are equivalent. Topologists use two natural numbers to keep track of what is called the *homology class* of the loops, the number of times they wrap the torus in the two directions.

What would happen if we were to mark a grid on a torus, and stitch hitomezashi? Interesting phenomena arise, that do not occur on the plane. There can't be lines of stitching that reach an edge, because there are no edges—all stitches are part of a loop. Patterns that formed bi-infinite paths on the plane—like jōkaku or yamagata— can form a loop around the torus body or hole. Dan tsunagi can loop multiple times around the hole. If there's an odd number of grid squares in one or both directions, the pattern and its dual get mixed together! Ren and Zhang (2024) have studied hitomezashi loops on the torus, and have established rules in terms of the homology class and the dimensions of the grid for the number of possible hitomezashi loops and the lengths of these loops.

If you'd like to stitch hitomezashi on a torus other than an idealised flat one (like those in Figure 11.10) you will run into some problems to solve. Unfortunately, while you can create a tube without introducing wrinkles, you can't simply sew a torus from a rectangle of even-weave fabric by closing up the ends of this tube. In three-dimensional space, the curvature of the physical torus requires us to cut or stretch the fabric we use to make it. bel castro (2008) has solved the shaping problem nicely in knitting—but to stitch hitomezashi on a smooth physical torus requires a grid spacing that varies throughout the surface.

11.6 OFF THE SQUARE GRID (AND INTO THE UNKNOWN)

There are a number of ways that we can take our stitching off the square grid whilst remaining on the plane. We can add diagonal stitches, as seen in Figure 11.11. The diagonal stitches have length $\sqrt{2}$, and by adding them in one direction only, all internal vertices have degree three. This traditional pattern, based on dan tsunagi is called *yabane* (arrow feather), resembling as it does the fletching of an arrow. Traditionally, diagonal lines of stitches are worked after the horizontal and vertical lines.

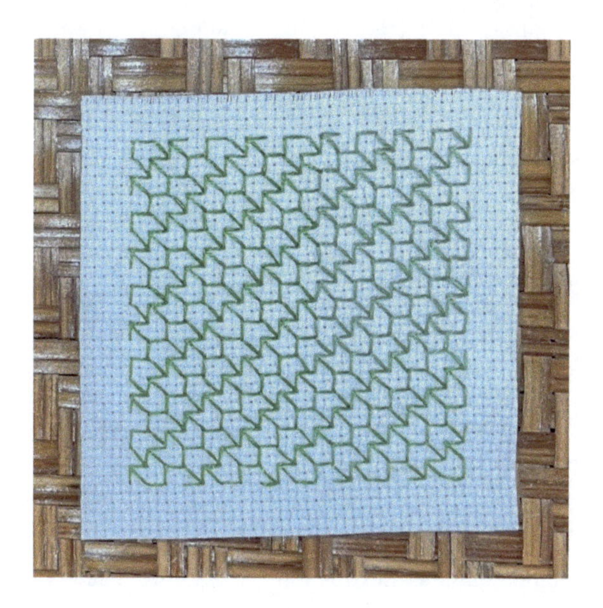

Figure 11.11 Hitomezashi stitch pattern incorporating diagonal lines: yabane. Stitching and photograph by the author.

Another consistent way to change stitch length is to make the stitches on the reverse a different length to those on the front. The two lengths used for the horizontal lines can be different from those used for the vertical lines. If this sounds a bit messy, it is true that stitches will cross one another between vertices, but Defant, Kravitz and Tenner (2023) have determined the possible types of patterns that can be achieved. They are in fact less interesting than the persimmon and snowflake loops of standard hitomezashi. They consist only of superposed dilated versions of kuchizashi, dan tsunagi and jōkaku.

We could take our hitomezashi onto a triangular grid, with the stitching forming degree three vertices (as Ayliean MacDonald did in her *Numberphile* video [Haran, 2021]; see also [Defant et al., 2023]). One possible way to improvise fabric on which to work such designs would be to mark a grid on plain fabric using a set square and soluble fabric marker. Another is to find a fabric printed with regular hexagons or equilateral triangles and use these as a stitching guide. (Yet another is to use the square grid plus one diagonal, as for yabane; the connectivity is the same, though the angles are not.) The pattern shown on the pocket of a shirt in Figure 11.12 is worked on the triangular lattice, but every second line

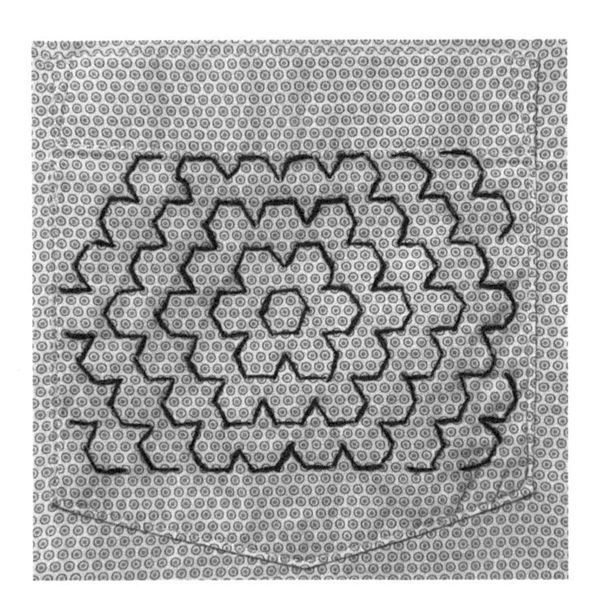

Figure 11.12 Hitomezashi stitching on a triangular grid provided by printed fabric, inspired by items posted to instagram by the Ōtsuchi Sashiko Recovery Project. Stitching and photograph by the author.

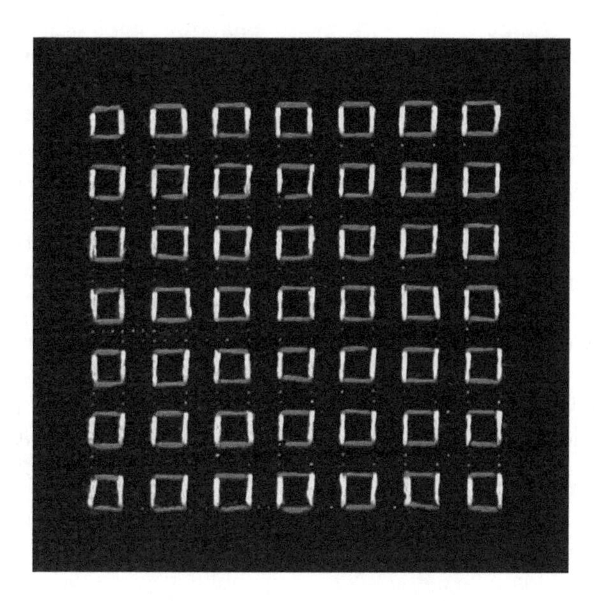

Figure 11.13 Two colours have been used to stitch kuchizashi. The colours are exchanged under rotation. Stitching and photograph by the author.

of stitching in each direction is empty. Each internal vertex has degree two; in two directions, there is a stitch in and a stitch out, and in the third there is no stitch. These designs seem simpler to understand than the degree three stitching. The motifs in Figure 11.12 suggest that at least some of the wallpaper patterns with three-fold or six-fold rotations (p311, p3m1, p31m, p611, p6mm) can be stitched as a hitomezashi variant.

Modern sashiko artists incorporate threads of more than one colour into the same design. In Figures 6.12 and 6.13, I used a second colour on a single section to highlight the stitching order. It's clear that this could be repeated at regular intervals and in both directions. Wallpaper patterns executed in more than one colour may have the property that the coloured sections change places under a transformation. For example, if all the horizontal lines of kuchizashi are stitched in red and all the vertical lines are stitched in white, as shown in Figure 11.13, the colours are exchanged when a 90° rotation is performed. There are 46 colour-exchanging wallpaper patterns, and 17 colour-exchanging frieze patterns. They are not so well known as the single colour versions, but can be found, for example, in the book by Washburn and Crowe (1988).

11.7 MORE TO EXPLORE

- Many geometric origami models, modular and made from a single sheet, exist. Experiment with making some from grid paper marked with hitomezashi designs. Can the design be kept correct at edges?
- Create more otedama stitched with hitomezashi patterns symmetric under the transformations comprising subgroups of C_{4v}.
- Could we stitch hitomezashi on a Möbius band? Why or why not?
- The patterns in Figure 11.12 suggest that wallpaper patterns with three-fold or six-fold rotations can be stitched as a hitomezashi variant. Can you design and stitch some?
- Explore colour-exchanging hitomezashi.

Tying off loosely

Throughout this book, we have limited our attention to hitomezashi formed from running stitch on a planar square grid, and insisted (with a few exceptions) that all stitches have the same length and colour, and that all internal vertices have degree two. These constraints have served us well across ten chapters, giving us many mathematical ideas to explore and beautiful stitching to create. Traditional and modern hitomezashi allow for variation from these rules, so there are ever more samplers to make and more chapters still to be written by curious and creative mathematicians and fibre artists.

Bibliography

[Ahmed, 2013] Ahmed, A. G. (2013). AA weaving. In *Bridges 2013 Conference Proceedings*, pages 263–270.

[Atiyah, 2007] Atiyah, M. (2007). Duality in mathematics and physics. https://fme.upc.edu/ca/arxius/butlleti-digital/riemann/071218_conferencia_atiyah-d_article.pdf

[AYUFISH int., 2023] AYUFISH int. (2023). *Amazing Sashiko.* Tuttle Publishing. (Translated from original Japanese edition of 2021).

[Baker and Goldstine, 2015] Baker, E. and Goldstine, S. (2015). *Crafting Conundrums.* CRC Press.

[belcastro, 2008] belcastro, s.-m. (2008). Only two knit stitches can create a torus. In s.-m. belcastro and Yackel, C., editors, *Making Mathematics with Needlework*, chapter 4, pages 54–68. A.K. Peters, Ltd.

[belcastro and Yackel, 2008] belcastro, s.-m. and Yackel, C., editors (2008). *Making Mathematics with Needlework.* A.K. Peters, Ltd.

[Bernasconi et al., 2008] Bernasconi, A., Bodei, C., and Pagli, L. (2008). On formal descriptions for knitting recursive patterns. *Journal of Mathematics and the Arts*, 2(1):9–27.

[Blondin Massé et al., 2009] Blondin Massé, A., Brlek, S., Garon, A., and Labbé, S. (2009). Christoffel and Fibonacci tiles. In *Discrete geometry for computer imagery*, volume 5810 of *Lecture Notes in Computer Science*, pages 67–78. Springer.

[Blondin Massé et al., 2010] Blondin Massé, A., Brlek, S., and Labbé, S. (2010). Combinatorial aspects of Escher tilings. In *22nd International Conference on Formal Power Series and Algebraic Combinatorics*, volume AN of *Discrete Mathematics and Theoretical Computer Science Proceedings*, pages 533–544.

[Blondin Massé et al., 2011] Blondin Massé, A., Brlek, S., Labbé, S., and Mendès France, M. (2011). Fibonacci snowflakes. *Annales des sciences mathematiques du Quebec*, 35(2):141–152.

[Blondin Massé et al., 2012] Blondin Massé, A., Brlek, S., Labbé, S., and Mendès France, M. (2012). Complexity of the Fibonacci snowflake. *Fractals*, 20(3-4):257–260.

[Bosch, 2012] Bosch, R. (2012). Truchet from Truchet tiles. https://gallery.bridgesmathart.org/exhibitions/2013-bridges-conference/bobb

[Brezine, 2004] Brezine, C. (2004). Creating symmetry on the loom. In Washburn, D. K. and Crowe, D. W., editors, *Symmetry Comes of Age: The Role of Pattern in Culture*, chapter 3, pages 65–80. University of Washington Press, Seattle.

[Briscoe, 2022] Briscoe, S. (2022). *Sashiko 365*. David and Charles, Exeter, UK.

[Campbell, 2022] Campbell, J. M. (2022). Artwork based on automatic sequences. *Journal of Mathematics and the Arts*, 16(4):287–308.

[Campochiaro, 2015] Campochiaro, C. (2015). *Sequence knitting: simple methods for creating complex fabrics.* Chroma Opaci Books.

[Ciletti and Pritelli, 2022] Ciletti, B. and Pritelli, M. C. (2022). *Sashiko.* Creative Editions.

[Clay, 2019] Clay, J. (2019). *Sashiko.* GMC Publications.

[Collins and Shepherd, 2013] Collins, J. and Shepherd, M. (2013). Botanica Mathematica. https://botanicamathematica.wordpress.com/about/

[Conway et al., 2008] Conway, J. H., Burgiel, H., and Goodman-Strauss, C. (2008). *The Symmetries of Things.* A. K. Peters Ltd.

[Crook, 2012] Crook, D. (2012). Remapping. https://annaschwartzgallery.com/exhibition/remapping

[Crowe, 2001] Crowe, D. (2001). Symmetries of culture. In *Bridges 2001 Conference Proceedings*, pages 1–20, Winfield, Kansas, USA.

[Cunliffe, 2022] Cunliffe, J. (2022). *Record, Map & Capture in Textile Art.* Batsford, London, UK.

[de Gier, 2009] de Gier, J. (2009). Fully packed loop models on finite geometries. In Guttman, A. J., editor, *Polygons, Polyominoes and Polycubes*, pages 317–346. Springer.

[Deane, 2012] Deane, A. R. (2012). Japanese garden reference guide. https://najga.org/handbook/. North American Japanese Garden Association.

[Defant and Kravitz, 2024] Defant, C. and Kravitz, N. (2024). Loops and regions in hitomezashi patterns. *Discrete Mathematics*, 347(1):113693.

[Defant et al., 2023] Defant, C., Kravitz, N., and Tenner, B. E. (2023). Extensions of hitomezashi patterns. *Discrete Mathematics*, 346(10):113555.

[Dietz, 1949] Dietz, A. K. (1949). *Algebraic Expansions in Handwoven Textiles.* The Little Loomhouse, Kentucky.

[Dow, 2007] Dow, M. (2007). Four knot. https://www.deviantart.com/markdow/art/Four-knot-1-63275749

[Franquemont and Franquemont, 2004] Franquemont, E. and Franquemont, C. R. (2004). Tanka, chong, kutij: Structure of the world through cloth. In Washburn, D. K. and Crowe, D. W., editors, *Symmetry Comes of Age: The Role of Pattern in Culture*, chapter 7, pages 177–214. University of Washington Press.

[Fukagawa and Horibe, 2014] Fukagawa, H. and Horibe, K. (2014). *Sangaku*-Japanese mathematics and art in 18th, 19th and 20th centuries. In *Bridges 2014 Conference Proceedings*, pages 111–118.

[Futatsuya, 2018] Futatsuya, A. (2018). Otsuchi recovery sashiko project. https://upcyclestitches.com/otsuchi-recovery-sashiko-project/

[Goldstine, 2016] Goldstine, S. (2016). Crystalline. *Knitty*, 57(Deep Fall 16).

[Goldstine, 2017] Goldstine, S. (2017). A survey of symmetry samplers. In *Bridges 2017 Conference Proceedings*, pages 103–110.

[Goldstine and Baker, 2011] Goldstine, S. and Baker, E. (2011). Crystallographic bracelet series. https://gallery.bridgesmathart.org/exhibitions/2012-joint-mathematics-meetings/ellie-baker

[Goldstine and Baker, 2012] Goldstine, S. and Baker, E. (2012). Building a better bracelet: wallpaper patterns in bead crochet. *Journal of Mathematics and the Arts*, 6(1):5–17.

[Goldstine and Yackel, 2022] Goldstine, S. and Yackel, C. (2022). A mathematical analysis of mosaic knitting: constraints, combinatorics, and colour-swapping symmetries. *Journal of Mathematics and the Arts*, 16(3):183–217.

[Gradit and Van Dongen, 2023] Gradit, P. and Van Dongen, V. (2023). A Truchet tiling hidden in Ammann-Beenker tiling. In *Bridges 2023 Conference Proceedings*, pages 69–76.

[Griffin, 1987] Griffin, M. (1987). Wear your own theory! *New Scientist*, 1987(March 26).

[Grovier, 2022] Grovier, J. (2022). Janet Sobel: The woman written out of history. https://www.bbc.com/culture/article/20220307-janet-sobel-the-woman-written-out-of-history. (BBC Culture).

[Haahr, 2023] Haahr, M. (1998–2023). RANDOM.ORG: True Random Number Service. https://www.random.org

[Haran, 2021] Haran, B. (2021). Hitomezashi stitch patterns. https://www.youtube.com/watch?v=JbfhzlMk2eY. Presented by Ayliean McDonald.

[Hargittai and Lengyel, 1984] Hargittai, I. and Lengyel, G. (1984). The seven one-dimensional space-group symmetries illustrated by Hungarian folk needlework. *Journal of Chemical Education*, 61(12):1033–1034.

[Hargittai and Lengyel, 1985] Hargittai, I. and Lengyel, G. (1985). The seventeen two-dimensional space-group symmetries in Hungarian needlework. *Journal of Chemical Education*, 62(1):35–36.

[Hayes, 2019] Hayes, C. (2019). Sashiko needlework reborn: From functional technology to decorative art. *Japanese Studies*, 39(2):263–280.

[Henderson, 1991] Henderson, E. (1991). Otedama—a fading Japanese juggling tradition. *Juggler's World*, 43).

[Holden, 2008] Holden, J. (2008). The graph theory of blackwork embroidery. In s.-m. belcastro and Yackel, C., editors, *Making Mathematics with Needlework*, chapter 9, pages 136–153. A.K. Peters, Ltd.

[Holden and Holden, 2021] Holden, J. and Holden, L. (2021). A survey of cellular automata in fiber arts. In Sriraman, B., editor, *Handbook of the Mathematics of the Arts and Sciences*, pages 443–465. Springer International Publishing.

[Hooper, 2013] Hooper, W. P. (2013). Renormalization of polygon exchange maps arising from corner percolation. *Inventiones Mathematicae*, 191(2):255–320.

[Hooper, 2019] Hooper, W. P. (2019). A curve in a quasi-periodic Truchet tiling. https://gallery.bridgesmathart.org/exhibitions/2019-icerm-illustrating-mathematics/wphooper

[Horiuchi, 2023] Horiuchi, H. (2023). *Mending with Boro.* Tuttle Publishing.

[Howard, 2017] Howard, K. (2017). Knit one, compute one. http://opentranscripts.org/transcript/knit-one-compute-one/

[Iiduka, 2021] Iiduka, S. (2021). *Sashiko for Making and Mending.* Tuttle Publishing. Translated from the Japanese edition of 2019.

[Irvine, 2015] Irvine, V. (2015). Lace orbifold *442. https://gallery.bridgesmathart.org/exhibitions/2015- bridges-conference/virvine

[Irvine et al., 2020] Irvine, V., Biedl, T., and Kaplan, C. S. (2020). Quasiperiodic bobbin lace patterns. *Journal of Mathematics and the Arts*, 14(3):177–198.

[Irvine and Ruskey, 2014] Irvine, V. and Ruskey, F. (2014). Developing a mathematical model for bobbin lace. *Journal of Mathematics and the Arts*, 8(3-4):95–110.

[Irvine and Ruskey, 2017] Irvine, V. and Ruskey, F. (2017). Aspects of symmetry in bobbin lace. In *Bridges 2017 Conference Proceedings*, pages 205–212.

[Janniere, 1994] Janniere, J. (1994). Filling in quilt history: A 16th century French patchwork banner. *The Quilt Journal–An International Review*, 3(1):1–6.

[Jensen, 2023] Jensen, S. (2023). Sequence knitting. *Journal of Mathematics and the Arts*, 17(1-2):111–139.

[Jones, 1991] Jones, K. (1991). Dicing with Mozart. *New Scientist*, 132(1799):26–29.

[Kaplan, 2019] Kaplan, C. S. (2019). Hexagonal cross stitch. https://isohedral.ca/hexagonal-cross-stitch/

[Keene, 1967] Keene, D. (1967). *Essays in idleness:The Tsurezuregusa of Kenkō.* Columbia University Press, New York. Translation from original Japanese by Keene.

[Kennedy, 2011] Kennedy, C. (2011). Binaries. In *The Taste of River Water*, page 22. Scribe.

[Kitagawa and Revell, 2023] Kitagawa, K. and Revell, T. (2023). *The Secret Lives of Numbers.* Penguin.

[Kitigawa, 2021] Kitigawa, T. L. (2021). Book illustration and the development of Japanese mathematics in the 1620s. *Journal of Mathematics and the Arts*, 15(1):33–53.

[Kitigawa, 2022] Kitigawa, T. L. (2022). The origin of Bernoulli numbers: Mathematics in Basel (Switzerland) and Edo (Japan) in the early 18th century. *The Mathematical Intelligencer*, 44(1):46–56.

[Koss, 2021] Koss, L. (2021). One-color frieze patterns in friendship bracelets: A cross-cultural comparison. In *Bridges 2021 Conference Proceedings*, pages 253–256.

[Labbé, 2010] Labbé, S. (2010). Fibonacci tiles can appear in a fully packed loop diagram. http://www.slabbe.org/blogue/2010/10/fibonacci-tiles-can-appear-in-a-fully-packed-loop-diagram/

[Lawrence, 2018] Lawrence, C. (2018). Play Truchet: Using the Truchet tiling to engage the public with mathematics. In *Bridges 2018 Conference Proceedings*, pages 359–362.

[Marchand et al., 2022] Marchand, R., Marcovici, I., and Siest, P. (2022). Corner percolation with preferential directions. https://arxiv.org/abs/2212.04399

[Marquez, 2018] Marquez, J. (2018). *Make and Mend.* Watson Guptill.

[Matsumoto et al., 2018] Matsumoto, E., Segerman, H., and Serriere, F. (2018). Möbius cellular automata scarves. https://gallery.bridgesmathart.org/xhibitions/2018-bridges-conference/mobscarves

[McKim, 1962] McKim, R. S. (1962). *101 Patchwork Patterns.* Dover Publications Inc. Revised version of work originally published 1931 by McKim Studios.

[Mitchell, 2016] Mitchell, D. (2016). The Robert Neale octahedron. https://www.origamiheaven.com/pdfs/nealesoctahedron.pdf

[Mitchell, 2024] Mitchell, D. (2024). Neale's skeletal octahedron. https://www.origamiheaven.com/historynealesoctahedron.htm. The Public Paperfolding History Project.

[Monnerot-Dumaine, 2009] Monnerot-Dumaine, A. (2009). The Fibonacci word fractal. https://hal.science/hal-00367972/fr/

[Mui, 2016] Mui, W. L. (2016). Simple symmetric symmetry sampler. https://gallery.bridgesmathart.org/exhibitions/2016-joint-mathematics-meetings/winglmui

[Museum of Modern Art, 2013] Museum of Modern Art (2013). MoMA's Jackson Pollock conservation project: Insight into the artist's process. https://www.moma.org/explore/inside_out/2013/04/17/momas-jackson-pollock-conservation-project-insight-into-the-artists-process/

[Nihon Vogue, 2020] Nihon Vogue (2020). *Simply Sashiko.* Tuttle Publishing. Translated from Japanese edition of 2019.

[OEIS Foundation Inc., 2024] OEIS Foundation Inc. (2024). The on-line encyclopedia of integer sequences. https://oeis.org

[Perkins and Seaton, 2020] Perkins, A. and Seaton, K. A. (2020). #MathArtChallenge Day 14: Hitomezashi stitching. https://arbitrarilyclose.com/2020/03/29/ mathartchallenge-day-14-hitomezashi-stitching-suggested-by-katherine-seaton/

[Pete, 2008] Pete, G. (2008). Corner percolation on $\mathbb{Z}^2$ and the square root of 17. *The Annals of Probability*, 36(5):1711–1747.

[Ponting, 2016] Ponting, A. (2016). Thue-Morse art. https://www.adamponting.com/thue-morse-art/

[Ramírez et al., 2014] Ramírez, J. L., Rubiano, G. N., and De Castro, R. (2014). A generalization of the Fibonacci word fractal and the Fibonacci snowflake. *Theoretical Computer Science*, 528:40–56.

[Reimann, 2023] Reimann, D. A. (2023). Connections between hitomezashi patterns and truchet tiling. In *Bridges 2023 Conference Proceedings*, pages 457–460.

[Ren and Zhang, 2024a] Ren, Q. and Zhang, S. (2024a). A succint proof of Defant and Kravitz's theorem on the length of hitomezashi loops. *Annals of Combinatorics*. Not yet assigned to an issue.

[Ren and Zhang, 2024b] Ren, Q. and Zhang, S. (2024b). Toroidal hitomezashi patterns. https://arxiv.org/pdf/2309.02741

[Roe, 2004] Roe, P. G. (2004). At play in the fields of symmetry. In Washburn, D. K. and Crowe, D. W., editors, *Symmetry Comes of Age: The Role of Pattern in Culture*, chapter 9, pages 232–303. University of Washington Press.

[Rothman and Fukagawa, 1998] Rothman, T. and Fukagawa, H. (1998). Japanese temple geometry. *Scientific American*, 278:84–91.

[Seaton, 2021] Seaton, K. A. (2021). Textile D-forms and D_{4d}. *Journal of Mathematics and the Arts*, 15(3-4):207–217.

[Seaton and Hayes, 2023] Seaton, K. A. and Hayes, C. (2023). Mathematical specification of hitomezashi designs. *Journal of Mathematics and the Arts*, 17(1-2):156–177.

[Shaver, 1992] Shaver, C. (1992). Sashiko: A stitchery of Japan. *Textile Society of America Symposium Proceedings*, 584:99–103.

[Shepherd, 2008] Shepherd, M. (2008). Symmetry patterns in cross-stitch. In s.-m. belcastro and Yackel, C., editors, *Making Mathematics with Needlework*, chapter 5, pages 70–89. A.K. Peters, Ltd.

[Shepherd, 2022] Shepherd, M. (2022). Knitting Life. https://www.madeleineshepherd.co.uk/gallery_779024.html

[Shepherd, 2018] Shepherd, M. D. (2018). Variations on Snake Trail quilting patterns. In Yackel, C. and s.-m. belcastro, editors, *Figuring Fibres*, chapter 4, pages 82–99. The American Mathematical Society.

[Sigler, 2003] Sigler, L. (2003). *Fibonacci's Liber Abaci: A Translation into Modern English of Leonardo Pisano's Book of Calculation.* Springer New York.

[Sire et al., 1989] Sire, C., Mosseri, R., and Sadoc, J.-F. (1989). Geometric study of a 2D tiling realted to the octagonal quasiperiodic tiling. *Journal de Physique*, 50(24):3463–3476.

[Smith and Boucher, 1987] Smith, C. S. and Boucher, P. (1987). The tiling patterns of Sebastien Truchet and the topology of structural hierarchy. *Leonardo*, 20(4).

[Spencer, 2018] Spencer, S. (2018). Networked knitting machine: not your average knit one, purl one.
https://www.raspberrypi.com/news/knitting-network-printer/

[Steckles, 2020] Steckles, K. (2020). Aperiodical design competition: Fractal bunting. https://aperiodical.com/2020/09/aperiodical-design-competition-fractal-bunting/.

[Stewart, 2015] Stewart, I. (2015). In the lap of the gods. In Brooks, M., editor, *Chance*, pages 97–107. Profile Books.

[Taimiņa, 2009] Taimiņa, D. (2009). *Crocheting Adventures on Hyperbolic Planes.* A.K. Peters Ltd.

[Truchet, 1704] Truchet, S. (1704). Memoir sur les Combinaisons. *Memoires de l'Académie Royale des Sciences*, pages 363–372.

[Uemae, 1997] Uemae, C. (1997). Chiyu Uemae. In Kournis, M., editor, *Art Textiles of the World—Japan*, pages 92–101. Telos Art Publishing.

[Vernier et al., 2016] Vernier, E., Jacobsen, J. L., and Saleur, H. (2016). Dilute oriented loop models. *Journal of Physics A: Mathematical and Theoretical*, 49(6):064002.

[Vincentelli, 2011] Vincentelli, M. (2011). Japanese sashiko textiles. *The Journal of Modern Craft*, 4:99–103.

[Washburn and Crowe, 1988] Washburn, D. K. and Crowe, D. W. (1988). *Symmetries of Culture: Theory and Practice of Plane Pattern Analysis.* University of Washington Press, Seattle.

[Washburn and Crowe, 2004] Washburn, D. K. and Crowe, D. W., editors (2004). *Symmetry Comes of Age: The Role of Pattern in Culture.* University of Washington Press, Seattle.

[Wildstrom, 2008] Wildstrom, D. J. (2008). The Sierpiński variations: Self-similar crochet. In s.-m. belcastro and Yackel, C., editors, *Making Mathematics with Needlework*, chapter 3, pages 40–52. A.K. Peters, Ltd.

[Yackel, 2011] Yackel, C. (2011). Spherical symmetries of temari. In s.-m. belcastro and Yackel, C., editors, *Crafting by Concepts*, chapter 8, pages 151–185. A. K. Peters, Ltd.

[Yackel, 2021] Yackel, C. A. (2021). Wallpaper patterns admissible in itajime shibori. *Journal of Mathematics and the Arts*, 15(3-4):232–244.

[Zeleny, 2013] Zeleny, E. (2013). Ada Deitz Polynomials for Handwoven Textiles. http://demonstrations.wolfram.com/AdaDeitzPolynomials ForHandwovenTextiles/

Index